INSTANT EGGHEAD GUIDE: THE UNIVERSE

J R MINKEL AND

SCIENTIFIC AMERICAN

WITH A FOREWORD BY GEORGE MUSSER JR.

INSTANT EGGHEAD

GUIDE: THE UNIVERSE

ST. MARTIN'S GRIFFIN ✈ NEW YORK

 For information, address St. Martin's Press, 175 Fifth Avenue, New York, N.Y. 10010.

www.stmartins.com

Library of Congress Cataloging-in-Publication Data

Minkel, J. R.
Instant egghead guide : the universe / JR Minkel & Scientific American ; with a foreword by George Musser Jr.—1st ed.
p. cm.
ISBN-13: 978-0-312-38637-5
ISBN-10: 0-312-38637-0
1. Astronomy—Popular works. 2. Astrophysics—Popular works. I. Scientific American, inc. II. Title. III. Title: Universe.
QB44.3.M56 2009
520—dc22
2009010883

First Edition: August 2009

10 9 8 7 6 5 4 3 2 1

To my parents

CONTENTS

ACKNOWLEDGMENTS

JR Minkel thanks Tom Weiler, Professor of Physics at Vanderbilt University, and Jonathan Pritchard of Harvard College Observatory, for checking portions of the manuscript for accuracy. Any remaining errors are the author's.

FOREWORD

by George Musser Jr.

Please do not drop this book into a time machine that is travelling back to the nineteenth century. It will only confuse them. The world is made of tiny particles? The sun is a big nuclear reactor? The passage of time can slow down? The universe is growing? Any scientists who found this book lying on the ground outside the time portal would have been baffled and humbled—assuming they didn't simply dismiss it out of hand as fraudulent, too fantastic to possibly be true.

The world has changed considerably in the past century, but hardly more so than in the discoveries made by modern science. People often talk about the wondrous inventions that science has made possible, from airplanes to electric toothbrushes, but even more significant is the new perspective science has given us on the world. It has enlarged our thinking and shown the beautiful variety and complexity of what is all around us. Many of the discoveries of the past century had been germinating long before then, but it took new experimental and conceptual breakthroughs for them to click into place.

This book manages the impressive feat of breaking these

ideas down into pieces that don't make huge demands on your time—in fact, I found that they fit neatly in between subway stops. Its step-by-step approach is not only convenient, but also conveys the way that science is cumulative, slowly building on its previous achievements. And the author JR Minkel's wry humor reflects most scientists' attitudes toward their work. They seek to understand the world for the pleasure of it and figure it doesn't hurt to have some fun along the way.

A century (or maybe even just a few years) from now, people may look back and see a new scientific revolution germinating from dimly recognized connections among the ideas encapsulated in this book, not to mention discoveries that have yet to be made. Aspects of the world that seem strange to us now will start to make sense. Others may make even less sense—which is good, because it keeps life interesting. So if a book from the future falls out of a time machine at your feet and it sounds fantastic, please don't be tempted just to throw it away.

George Musser Jr. is the astronomy editor at *Scientific American* magazine and author of *The Complete Idiot's Guide to String Theory*.

CHAPTER

ONE

MATTER AND ENERGY

ELECTRONS, PROTONS, AND NEUTRONS

THE BASICS

The world, as any egghead knows, is made of atoms. And the raw ingredients of atoms are called subatomic particles. We encounter a number of particles in this book, but to understand matter, from the book in your hands to the core of a star, we need a good grasp of these three: electrons, protons, and neutrons. So let's get to know them.

The electron is an elementary particle, which means it can't be broken down into other particles. It is very tiny—it may have no size at all, in fact—and has very little mass. (All matter has mass, which is a measure of the oomph it takes to get something moving.) An electron carries a negative electric charge, which is the opposite of a positive charge. Like charges repel; opposite charges attract.

Unlike electrons, protons and neutrons are not elementary particles; they are made of smaller particles called quarks, which we get to later. Protons are positively charged, and neutrons are electrically neutral. Protons and electrons pair up due to their electric attraction but the proton dominates the relationship because it has the mass of some 1,800 electrons. Neutrons are slightly heavier than protons but otherwise identical.

ON THE FRONTIER

Everything we see, including Earth and the stars, is made of atoms, but as it turns out, there's a lot of additional matter we can't see. Scientists call it dark matter, and it's completely invisible. The only reason we think it's there is because of its effects on the other stuff in the universe. Figuring out the identity of dark matter is one of the biggest challenges in science today. But let's put that on the shelf for the moment. Studying the behavior of ordinary matter will take us a long way toward understanding how the universe came to be the way it is.

COCKTAIL PARTY TIDBITS

- All subatomic particles of a single type are identical. It's impossible to tag one of them the way you would tag a penguin or a seal to study its migratory patterns. So if you discover something about one particle, it applies to all the rest. It's a law of nature!
- The charge of an electron is one of the fundamental physical constants. We can't explain it from more basic principles; we can only measure it.
- American physicists Robert Millikan and Harvey Fletcher were the first to measure the charge of an electron in 1909 by suspending droplets of oil in an electric field. (They were slightly off but, hey even eggheads make mistakes.)

ATOMS

THE BASICS

The structure of an atom is a bit like the solar system. At the center is a dense nucleus made of protons and neutrons. Electrons orbit the nucleus in a complicated way that is more like a cloud than like planets. (But more on that later.) The nucleus is positively charged, so it attracts electrons to it, one for each proton. As long as the number of protons and electrons is the same, the atom is electrically neutral, which is good, because otherwise we'd be walking around shooting sparks everywhere. If an atom gains or loses electrons, say from friction or heat, it becomes electrically charged and we call it an ion.

Atoms come in 94 naturally occurring types, called elements, distinguished by the number of protons in the nucleus, called the atomic number. Hydrogen, with a single proton, is the lightest element. Because electrons are so light, more than 99.9 percent of an element's mass comes from its protons and neutrons. All elements exist in multiple forms with slightly different masses, called isotopes. The difference is the number of neutrons in the nucleus.

Some isotopes are unstable; they break down into other elements in a process called radioactive decay. All natural elements have radioactive isotopes mixed in with them. Researchers can estimate the age of fossils, space rocks, and other ancient samples by comparing the ratio of isotopes in the specimen.

ON THE FRONTIER

Because atoms are so tiny, we should be forgiven for needing thousands of years to prove their existence. In the nineteenth century, scientists found they could explain the behavior of gases and liquids by assuming that atoms knocked one another around like billiard balls. Today we can detect atoms directly thanks to special tools such as the electron microscope, which scans surfaces using a thin beam of electrons.

In 2008, University of California, Berkeley, researchers boosted the sensitivity of electron microscopy enough to pick out single hydrogen atoms—the lightest of all—suspended on an extremely flat surface.

COCKTAIL PARTY TIDBITS

- **If you could enlarge an apple to the size of Earth, the atoms inside it would be as big as the original apple. Bite on that!**
- **The idea of atoms dates back thousands of years to early eggheads such as Democritus of Greece, who argued that matter must be made of particles that can't be split into smaller pieces. (The word *atom* comes from the Greek for "uncuttable.") They were half right. Atoms are the smallest units of elements, not matter.**

THE ELEMENTS

THE BASICS

Elements are substances that can't be broken down into simpler substances. We classify the elements based on their location in the periodic table of elements, bequeathed to schoolchildren everywhere by Russian chemist Dmitri Mendeleev in 1869. Today we know they represent different types of atoms.

The modern periodic table is arranged in rows and columns. Elements in the same column have similar chemical properties. For example, the alkali metals—lithium, sodium, potassium, and so on—are so highly reactive that they explode on contact with water; the noble gases—helium, neon, argon, and so on—are all inert, meaning they resist forming molecules.

The rows are trickier. Electrons orbit the nucleus in different regions called shells, which can fill up as seats do around a table. An atom likes to have a full shell. The chemistry of an element depends on how close its outermost shell is to full. If an atom has one electron too few or too many, it may just grab another electron (or give away its own) and become an ion.

Some elements are common, others very rare. If you can name an element, it's probably common. Earth and other rocky planets are made of silicon, iron, carbon, nitrogen, phosphorous, and a host of less common elements. Earth's atmosphere consists mainly of nitrogen and oxygen.

ON THE FRONTIER

Elements heavier than fermium (100 protons) are generally unstable, lasting anywhere from days to hours to a few seconds or less. In 2006, a team from Lawrence Livermore National Laboratory in Berkeley synthesized superheavy element 118 by colliding isotopes of californium (98 protons) and calcium (20 protons). It decayed into lighter elements in .9 milliseconds.

The periodic table proved its virtues yet again in 2007, when Swiss researchers reported that short-lived, superheavy element 112 ("ununbium") formed bonds with gold atoms in the same way as zinc and mercury, its column-mates on the table.

COCKTAIL PARTY TIDBITS

- **The most abundant elements in the universe are hydrogen and helium.**
- **You can actually buy most of the elements online, even some of the radioactive ones. There are people who collect elements as a hobby.**
- **The elements Earth is made of (including those in our bodies) were born about five billion years ago inside dying stars that exploded and spread their ashes into space.**

MOLECULES

THE BASICS

A molecule is a combination of elements bonded together into one unit. When two atoms don't have enough electrons individually to fill their outer orbitals, they can achieve fullness by pooling electrons, like pushing two tables together in a restaurant. A hydrogen atom has one electron but needs two to be full. If two hydrogen atoms share their two electrons, both are happy. The sharing of electrons is known as a covalent bond.

Molecules made of two or more elements are called compounds. Some famous examples include water, which is made of two hydrogen atoms bonded to an oxygen atom (abbreviated H_2O) and carbon dioxide (CO_2). Atoms of the same element can also form molecules, such as hydrogen (H_2), oxygen (O_2), and ozone (O_3). Those elements that are blessed with fullness (helium is one) don't like to make molecules of any kind. In other words, they're chemically inert.

Molecules are never perfectly electrically neutral. Some atoms in a molecule may hog the shared electrons. An example is the oxygen atom in water. The electrons stick closer to oxygen than they do to the hydrogen atoms, which have smaller nuclei. So the oxygen has a partial negative charge, and the hydrogen atoms have partial positive charges. As a result, water molecules tend to stick together by lining up their positive and negative ends.

ON THE FRONTIER

Among the elements, one of the best sharers is carbon. It needs eight electrons in its outer shell but it only has four. So it likes to make four bonds, typically with other carbon atoms but also with hydrogen, oxygen, nitrogen, and other elements. There's a whole science of carbon sharing, called organic chemistry. Life on Earth is made of carbon molecules, also called organic molecules. We eat them, break them down, and build new ones to run our bodies. When searching for alien life, scientists think we should be looking for signs of carbon chemistry, or at least the possibility of it.

COCKTAIL PARTY TIDBITS

- **Not all compounds are molecules. When two ions are joined together, it's called an ionic compound or a salt. Table salt is made of positively charged sodium and negatively charged chlorine.**
- **Wall-climbing geckos have evolved to take advantage of a weaker kind of intermolecular bonding. Their feet are covered in millions of bristlelike setae, designed to maximize the Van der Waals interaction, which occurs when electrons on adjacent molecules twitch back and forth simultaneously.**
- **In one of Albert Einstein's most cited papers, he calculated the size of molecules by analyzing Brownian motion, the zigzag path of pollen and other tiny grains in liquid.**

CHEMICAL ENERGY

THE BASICS

It takes energy to break chemical bonds. But like relationships, some molecules are easier to break apart than others. Pump enough heat into any molecule and its atoms will split and go their separate ways. That's why Bunsen burners come in so handy in chemistry class. Freed from their molecular shackles, elements can reshuffle themselves to form new, more stable molecules. This process is called a chemical reaction.

Chemical bonds are a source of energy. In an exothermic reaction, broken bonds release their energy as heat. This is what happens when you start a fire or run an engine. Heat from friction, a match, or a spark plug splits apart some hydrocarbon molecules, which release their energy as heat, which splits apart more molecules and so on.

The same basic thing happens in our bodies. We eat hydrocarbon molecules (sugars and fats) and the body uses specially shaped molecules called enzymes to break them apart. The energy released by the broken bonds is channeled in complicated ways to do everything from moving our muscles, growing and repairing our cells, to powering our brains. (See the *Instant Egghead Guide: The Mind* for more on what happens in that thing.)

ON THE FRONTIER

We get most of our electrical power from burning fossil fuels. But there are other ways to extract chemical energy, such as the fuel cell, a device that generates a current by passing protons (aka hydrogen ions) through a membrane and fusing them with oxygen to make water. Not coincidentally, this is similar to how our bodies use oxygen. Our cells strip high-energy protons from hydrocarbon molecules, use them to make a kind of current to power the cells, and then combine them with oxygen to make water. Spent carbon molecules are exhaled as CO_2.

COCKTAIL PARTY TIDBITS

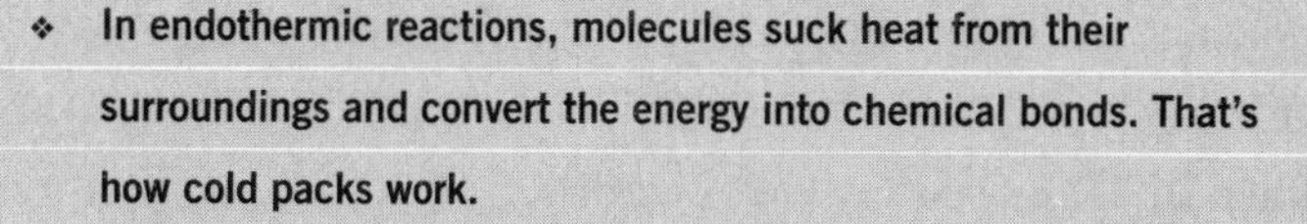

- In endothermic reactions, molecules suck heat from their surroundings and convert the energy into chemical bonds. That's how cold packs work.
- Exothermic reactions are useful for blowing things up. Breaking the bonds in molecules of tri-nitro-toluene (TNT) will release a lot of energy very quickly on their surroundings.
- The atoms in a molecule are constantly vibrating. Heat drives chemical reactions because it makes atoms vibrate more.

STATES OF MATTER

THE BASICS

Here on Earth, every material—elemental or compound—comes in one of three states: solid, liquid, and gas. States of matter exist because molecules stick to one another via weak chemical bonds such as those between water molecules. Those links give them properties that they don't have as individual particles, such as hardness or wetness. The temperature at which a substance melts (or boils) depends on the strength of its chemical ties. Molecules in rock are fused together much tighter than those of water.

Molecules are constantly jiggling, which we call heat or temperature (which are not quite the same thing, as we will see). In a solid, molecules aren't jiggling enough to overcome the links between them. They hold their shapes, like a bunch of LEGO bricks stuck together. Turn up the heat and the molecules will begin shaking so much they break free of their neighbors and start to mill around like when you poke around in a tray of loose LEGO bricks. The solid has become a liquid, which flows but doesn't change volume.

In a gas, the bonds between molecules have broken entirely. They float around with no overall shape and their volume expands with temperature. A hot gas is less dense than a cold gas. That's why hot air rises.

ON THE FRONTIER

When a gas gets very hot—we're talking thousands or millions of degrees—heat vibrations shred some of the atoms apart into electrons and nuclei. This electrically charged cloud is called plasma. It's what the sun and stars are made of, which means it's the most common state of matter in the universe. Plasma-screen TVs work by generating a plasma that emits light. On a larger scale, scientists are working on creating powerful magnetic fields to control plasma to make nuclear fusion reactions.

COCKTAIL PARTY TIDBITS

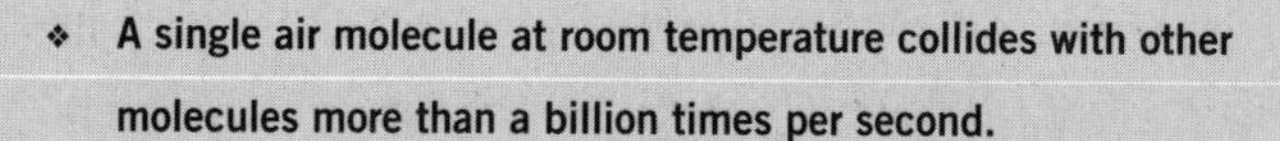

- A single air molecule at room temperature collides with other molecules more than a billion times per second.
- Microwave ovens work by generating electric fields that rapidly oscillate back and forth, causing atoms to vibrate.
- When helium is chilled to nearly –459 degrees Fahrenheit, it loses all viscosity and becomes a superfluid, capable of climbing up the sides of a container like some kind of liquid alien life form.

CONSERVATION OF ENERGY

THE BASICS

The word *energy* comes up a lot these days. But what is it? We know it comes in many forms. We eat food to give ourselves energy; we burn fossil fuels so electrical energy will come out of our wall outlets. To an egghead, the common denominator is that energy is the capacity to cause a change. It's a basic property of matter.

One of the most fundamental observations researchers have ever made is that no matter how energy is transformed, the amount you end up with is always the same as the amount with which you started. This is called conservation of energy.

The different forms of energy boil down to two basic types. Moving objects have kinetic energy, which they impart to whatever they bump into. When you walk, you add energy to the sidewalk by jostling the atoms beneath your feet. Sound energy is the kinetic energy of waves of molecules spreading outward like ripples in a pond.

Even when an object is at rest, it has the potential to cause a change. We say it has potential energy, which is the other basic type of energy. It comes from gravity and other forces. A glass of water in your hand has the potential to fall. It has gravitational potential energy. Chemical bonds store chemical energy.

ON THE FRONTIER

Essentially all the energy on Earth comes from the sun. Light from the sun heats the ground, the atmosphere, and our bare arms. Plants capture solar energy and turn it into hydrocarbons, which we eat or feed to animals so we can eat them. Fossil fuels are just ancient plants and animals squished to goo in Earth's crust.

We can tap the sun's energy directly using solar cells, materials that absorb light and turn it into electricity. In principle, they are a cleaner source of energy than oil and other fossil fuels, which produce the greenhouse gas carbon dioxide when we burn them. Other ways of tapping nonfossil fuel energy include windmills, waterfalls, and even heat from inside Earth.

COCKTAIL PARTY TIDBITS

- **The conversion of one type of energy to another can take many forms. Researchers have discovered that blasting liquids with ultrasound can generate bubbles that collapse so forcefully, they produce temperatures of millions of degrees.**
- **German surgeon Julius Robert von Mayer codiscovered the conservation of energy in 1842, after observing that his patients in the Dutch East Indies had redder blood than his usual patients, indicating they needed less oxygen to maintain their temperature.**

HEAT AND TEMPERATURE

THE BASICS

A very common form of energy is heat. We mentioned that heat and temperature are not the same thing. Temperature is the average amount of jiggling in a group of molecules. The water molecules in a mug of hot coffee are jiggling more violently than those in a glass of cold milk. Heat is temperature in motion. When you mix cold milk with hot coffee, the coffee molecules knock around the milk molecules until they are all jiggling by the same amount.

Heat can do work. As gas molecules vibrate, they bump into whatever's around them. Heating a gas makes the vibrations more violent, and the gas will expand if it's in a container that will let it. If not, its pressure rises, meaning the gas molecules knock into the container more frequently and violently. A car's engine works by burning air mixed with gasoline, which pushes a piston that ultimately turns the wheels.

No matter how carefully you operate a machine, you can never avoid wasting some energy as heat. It doesn't matter if the machine in question is a human-built device such as a car or a computer, an evolved organism such as a bacterium, or a denizen of the universe such as a star. Without a fresh infusion of energy now and then, it will eventually wind down and stop running. This idea is called the second law of thermodynamics.

ON THE FRONTIER

The light emitted from an object can tell you its temperature. Jiggling molecules emit infrared light, which our bodies sense as warmth. When the temperature of an object rises past a certain point, it emits visible light. Think of a glowing coal. The color of a heated object (the frequency of light it emits most intensely) depends only on its temperature. This is called blackbody radiation, and it's how we infer the temperature of the sun and other stars. Stars cool very slowly—over millions to trillions of years—because there's not much matter around them to be heated.

COCKTAIL PARTY TIDBITS

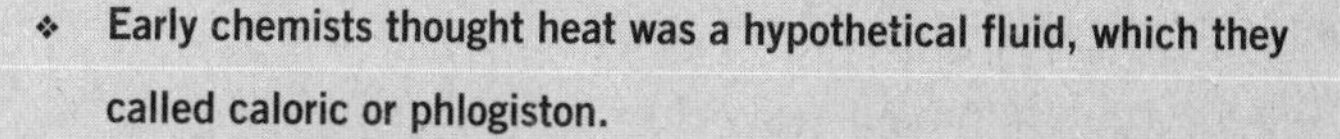

- Early chemists thought heat was a hypothetical fluid, which they called caloric or phlogiston.
- A rubber band makes for a snappy lesson in thermodynamics. When you stretch it, it heats up. (Try touching it to your lips.) You've converted mechanical energy to heat.
- The temperature of space comes from a faint glow of microwaves called the cosmic microwave background. If you put a thermometer in space, it would come to read 2.7 kelvins, or nearly –459 degrees Fahrenheit.

ENTROPY

THE BASICS

An implication of the second law of thermodynamics is that a system starved of fresh energy tends to become more "disordered" or mixed together over time. Entropy is a measure of how much mixing has taken place.

According to the second law, entropy can only increase or stay the same, never decrease, for any system that's isolated from its environment, be it the coffee in a thermos or a star in outer space. Differences in temperature "want" to smooth themselves out. Entropy tends to increase.

In the world of molecules, entropy is related to the number of different ways the same molecules can be mixed together. It's like having a messy room. There are many different ways for a room to be messy. The clothes can be hanging off the chair or on the floor; candy wrappers can be on the table or under the bed. But there are many fewer ways to have a clean room. The clothes have to be folded and hung up. The wrappers go in the wastebasket.

But as soon as you stop doing work, things start to get messy again. It's the same with molecules. Leave a bottle of perfume open, and perfume molecules will tend to jiggle themselves into the air. It's extremely unlikely they would all randomly jiggle back into the bottle.

ON THE FRONTIER

Entropy is related to the passage of time. The laws of physics don't seem to differentiate between going forward in time or going back. The only thing keeping vibrating perfume molecules from going back into the bottle is statistics. It's just way, way too unlikely. If we saw it happen, we'd say time was running in reverse. Scientists believe we experience the passage of time because the universe is gradually getting more disordered.

COCKTAIL PARTY TIDBITS

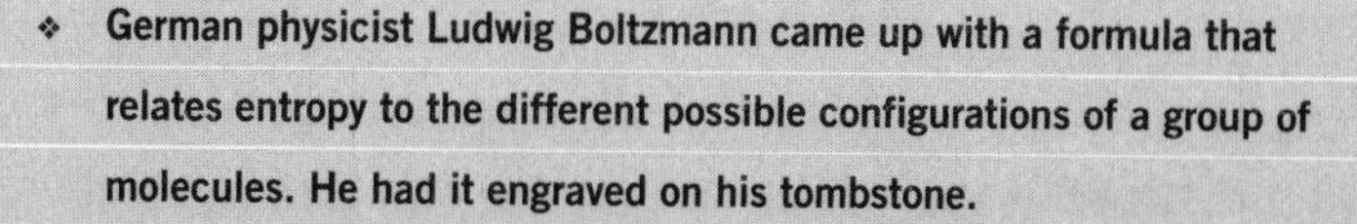

- **German physicist Ludwig Boltzmann came up with a formula that relates entropy to the different possible configurations of a group of molecules. He had it engraved on his tombstone.**
- **We said life doesn't violate the second law. Living things are definitely highly ordered, but they derive their order by creating more disorder around them than they contain within.**

THE NUCLEUS

THE BASICS

The nucleus is small, even by atomic standards. If a hydrogen nucleus were the size of a marble (about a centimeter across), its electron would lie 100 yards away. What could possibly be going on in a space that small, you might wonder. Just you wait. The nucleus is really the coolest part of the atom. Seemingly stable, it can shoot out pieces of itself, fuse with other nuclei, and even explode into showers of other particles.

Maybe you were wondering why the nucleus exists at all. If like charges repel, shouldn't the mutual electric repulsion of all those positively charged protons blow the nucleus apart? It turns out there's an even stronger force that counters the electric repulsion. It's called the strong force, and it makes protons and neutrons stick together like refrigerator magnets.

Protons packed too tightly? Just add neutrons and voilà: instant glue. Up to a point, anyway. If a nucleus has more than 83 protons, no amount of neutron glue can bind it together forever. It will eventually break down in a process called radioactive decay. A second nuclear force—the weak force—is responsible for one kind of radioactivity.

ON THE FRONTIER

We mentioned that the nucleus could melt. At the Relativistic Heavy Ion Collider (RHIC) in Long Island, New York, researchers smashed gold nuclei together at 99.99 percent light speed. When the nuclei struck one another, they melted into a cloud of quarks, particles inside protons and neutrons—and gluons, another kind of particle that "glues" quarks together, naturally enough. The mixture is called the quark–gluon plasma, and it's the hottest, densest matter scientists have ever made. Researchers believe the universe was full of quark–gluon plasma a few microseconds after the Big Bang.

COCKTAIL PARTY TIDBITS

- Incoming! English physicist Ernest Rutherford discovered the nucleus in 1909 by firing so-called alpha particles at gold foil. Light and positively charged, the alpha particles seemed to be striking small, positively charged nuggets; Rutherford likened it to cannon shells bouncing off tissue paper.
- The nucleus may also come in shells like electrons in an atom. Superheavy elements are thought to be stable because they have "magic" numbers of protons and neutrons that fill up their nuclear shells.
- The quark–gluon plasma has a temperature of trillions of degrees and a pressure of 10^{30} Earth atmospheres.

RADIOACTIVITY

THE BASICS

Radioactivity is a process that transforms one element into another. It occurs because the nucleus is under a lot of strain from the repulsion between protons and needs to relieve the pressure somehow. Elements heavier than bismuth (number 83 on the periodic table) are always radioactive, but all elements have radioactive isotopes.

When a radioactive isotope decays, it can gain or lose protons in one of two ways. In alpha decay, a helium nucleus (known as an alpha particle) quite literally burrows its way out of the nucleus, in a process called quantum tunneling. The alpha decay of uranium (element 92) yields thorium (element 90), for example. Beta decay is a little weirder. In it, a neutron morphs into a proton (via the weak force mentioned earlier) and emits an electron, called a beta particle. Two beta decays turn thorium back into a lighter isotope of uranium.

The intrinsically radioactive elements do a kind of alpha–beta shuffle—losing protons and then gaining them back—until they become lead, which has a stable nucleus. But it doesn't happen overnight. Radioactive samples decay at a rate called the element's half-life, the time it takes for half the atoms in a sample to break down. (The individual atoms decay at random.) The half-life of uranium is 4.5 billion years, which is about the age of Earth.

ON THE FRONTIER

Alpha and beta particles are hazardous because they can plow into our bodies and strip electrons from important molecules, potentially killing cells or causing mutations that might lead to cancer. Former Russian secret service agent Alexander Litvinenko was poisoned in 2006 in Great Britain using polonium-210, an alpha-emitting radioactive isotope with a half-life of 138 days. Little did his poisoners know that tests were sensitive enough to pick up traces of the isotope.

COCKTAIL PARTY TIDBITS

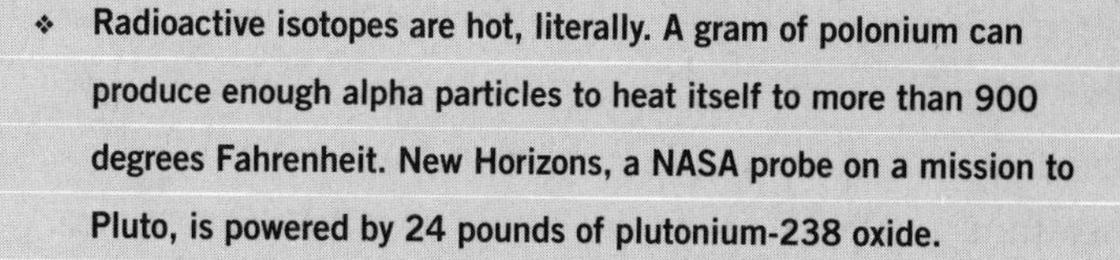

- Radioactive isotopes are hot, literally. A gram of polonium can produce enough alpha particles to heat itself to more than 900 degrees Fahrenheit. New Horizons, a NASA probe on a mission to Pluto, is powered by 24 pounds of plutonium-238 oxide.
- Neutrons are unstable outside of the nucleus. Left to themselves, they would decay (a fancy word for transform) into protons in about 15 minutes on average.
- Alpha and beta particles were named for their ability to penetrate matter. (Scientists didn't know their true identities.) A sheet of paper will block alpha particles but it takes an aluminum plate to halt beta particles.

FUSION

THE BASICS

When nuclei are crushed together and heated to millions of degrees they can fuse together to form heavier nuclei in a process called (duh) nuclear fusion. In the process, they liberate a tremendous amount of energy. The sun and other stars get their energy by fusing hydrogen into helium. Without fusion, the sun would not shine the way it does.

In the simplest fusion reaction, hydrogen nuclei fuse to form helium. As with a campfire, the reaction requires heat to get started. Protons resist coming together because of their electric repulsion. But when heated above 10 million degrees, they begin to rattle so violently that they touch, and once that happens, the nuclear forces kick in, turning protons into neutrons and linking them together into a helium nucleus.

In stars, it takes four protons to make helium through a series of reactions called the proton–proton cycle. The process is self-sustaining because it gives off a lot of energy. Where does the energy come from? A helium nucleus is 0.7 percent lighter than the sum of the masses of two protons and two neutrons. That missing mass has to go somewhere, and according to Einstein's equation $E=mc^2$, mass is equivalent to energy. The excess mass is converted to energy.

ON THE FRONTIER

Researchers would like to harness fusion to make electricity. Cooking matter to the temperatures required for fusion turns it to plasma, a hot gas of electrons and nuclei, which is dangerous and hard to contain. Gravity solves that problem in a star, but here on Earth we have to be more creative. Now an international team has finally begun work on a prototype fusion reactor called ITER, the International Thermonuclear Experimental Reactor, in southern France. In the works for more than 20 years, the $15 billion project is designed to generate a powerful doughnut-shaped magnetic field, called a tokamak, to heat and contain the plasma for sustained fusion reactions.

COCKTAIL PARTY TIDBITS

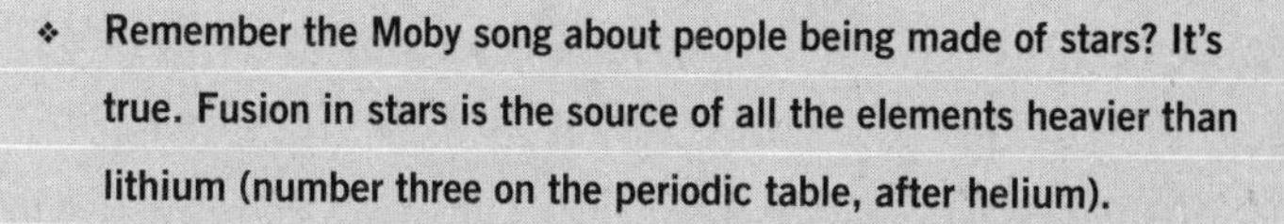

- **Remember the Moby song about people being made of stars? It's true. Fusion in stars is the source of all the elements heavier than lithium (number three on the periodic table, after helium).**
- **All the elements up through iron were made in the final week of a star's life. The elements from cobalt through uranium were made in the last moment before a star exploded in a supernova.**

CHAPTER

TWO

ELECTROMAGNETISM AND LIGHT

ELECTRICITY

THE BASICS

Electricity encompasses everything from sparks and lightning to the electrical energy coming from wall outlets and batteries to the static cling that makes your hair stand on end after rubbing it with a balloon. It's all about charged particles.

Around every charged particle is something called an electric field, a sort of invisible influence that causes like charges to repel each other and opposite charges to attract. The strength of the repulsion or attraction depends on how concentrated the charges are.

An electric current is a bunch of charged particles moving in a loop. Metals and other materials allow electrons to move around easily. They are called conductors. Materials such as glass and rubber are insulators; they resist the flow of electrons.

When you walk on a carpet wearing socks, atoms in the carpet pull electrons from the atoms in your socks. Because air is normally an insulator, the charges have nowhere to go and you are left slightly positively charged. Then, when you reach for a doorknob, electrons leap from the metal to your hand, creating a spark that gives you a shock.

The same thing happens in a thunderstorm. Clouds build up a positive charge until the electric field rips electrons from air molecules, allowing charged particles to flow between ground and sky. The current heats the air around it, producing the flash of lightning.

ON THE FRONTIER

Normally a conducting material is imperfect; it resists the flow of current by some nonnegligible amount. But when some conductors are chilled to extremely low temperatures, they suddenly become able to carry a current with almost zero resistance, a phenomenon called superconductivity. Because wires made of superconducting material can carry a lot more electricity than normal copper wires of the same size, some utility companies are installing them underground to boost power transmission in cities without ruining everyone's view with unsightly power lines.

COCKTAIL PARTY TIDBITS

- **The first documentation of static cling dates to 600 B.C., when Greek philosopher Thales of Miletus wrote about the effects of rubbing fur on substances such as amber. (The word *electricity* comes from the Greek for "amber".)**
- **In copper wire, electrons travel about a millimeter per second. Electric power is transmitted much faster than that because electrons push each other like water in a hose.**

MAGNETISM

THE BASICS

Magnetism is electricity's counterpart. Instead of charges, a magnet has two poles—north and south—which produce a magnetic field. As with electricity, opposite poles attract and like poles repel. Hence all the fun you can have playing with refrigerator magnets. Magnetic poles differ from charges because you can never isolate a single pole. If you break off a piece of a magnet, all you will find is a smaller magnet.

You can see the magnetic field around a magnet by sprinkling iron filings around it. The filings will line up in arcs spreading from either end of a bar magnet like a whisk. These are magnetic field lines, the units of a magnetic field. The magnet's field causes magnetic regions in the iron all to line up the same way, turning each filing into a little magnet that points the same way as the big magnet.

Moving charges create a magnetic field. This is the idea behind electromagnets: the electric current flowing through a wire wrapped around a nail or another piece of metal generates a magnetic field, which magnetizes the metal. To crank up the magnetic field, you boost the current or add more loops of wire.

The sun has a massive field that extends throughout the solar system, carrying charged particles called solar wind.

Earth's magnetic field is responsible for the aurora borealis, or Northern Lights.

ON THE FRONTIER

Here's how the supervillain Magneto could control your mind. In addition to being created by moving charges, magnetic fields in motion can generate a current. The right kind of magnetic field can influence electrical circuits in your brain, the basis of learning, memory, and thought (as far as we know). Researchers are investigating a technology called transcranial magnetic stimulation, which uses a strong magnet to scramble electric circuits in the area of the brain below the magnet. It's being evaluated for possible effects on migraines, Parkinson's disease, and clinical depression.

COCKTAIL PARTY TIDBITS

- **Electrons act as tiny magnets. A magnetic material is just one in which the magnetic poles of all the electrons add up in the right way.**
- **Researchers have used big magnets to levitate live frogs, grasshoppers, hazelnuts, tulips, and other organisms.**
- **Magnetic resonance imaging machines use strong magnetic fields to detect what's happening in the brain and body. The magnetic field makes hydrogen atoms wobble like little tops and emit radio waves.**

ELECTROMAGNETISM

THE BASICS

Electromagnetism is the double helix of physics. As the name implies, it ties together electricity and magnetism into a single force, which does everything from holding atoms and molecules together to making your cell phone work. Developed by Scottish physicist James Clerk Maxwell in the 1860s, electromagnetism is one of the great examples of unification in science, showing that two seemingly independent things are two sides of one coin.

Maxwell realized that electric and magnetic fields always come braided together, because any change in an electric field produces a changing magnetic field and vice versa. One of the most useful aspects of electromagnetism is induction, which causes a current in one wire to generate a current in a nearby wire via a magnetic field in between them. You've already seen it at work if you use an electric toothbrush that comes with a recharging station. It's the basis of modern electrical power.

Imagine that electric and magnetic fields are two Slinkies stretched out side by side. When an electron moves, it makes a ripple in the electric field like a Slinky given a shake. This electric ripple sets the magnetic Slinky in motion. The two vibrations feed on each other. If one shakes up and down, the other shakes side to side. Together, the paired vibrations zoom through empty space at the speed of light. In fact, they are light.

ON THE FRONTIER

Ever since Nikola Tesla, scientists have been looking for ways to broadcast electric power through the air like radio waves, but long-range induction typically involves things such as microwaves, which would cook us if we were bathed in them. In 2007, researchers at the Massachusetts Institute of Technology demonstrated they could transmit power to a 60-watt light bulb six feet away from a power source, by attaching a copper coil to the bulb and a matching one to the source. All without frying anyone.

COCKTAIL PARTY TIDBITS

- **Science fiction writer Arthur C. Clarke said sufficiently advanced technology is indistinguishable from magic. He must have had electromagnetism in mind when he said that.**
- **Some studies have found that people exposed to a lot of electromagnetic radiation from cell phones or power lines seem to get more cancer. In July 2008, the director of a Pittsburgh cancer center recommended his staff limit the time they spent with their cell phones against their ears, just in case.**

LIGHT WAVES

THE BASICS

Light is that bright stuff that comes from the sun and light bulbs. Maybe you've seen it? It carries energy from one place to another. Vision is only possible because molecules in our retinas absorb light energy streaming into the eyes from matter all around us.

In everyday terms, light is an electromagnetic wave, a ripple in the electromagnetic field. As any wave does, it has a speed, which depends on what it's moving through. The speed of light in water is slower than the speed of light in air. When you hear people talk about the speed of light, they mean its speed in a vacuum (empty space), which is 299,792,458 meters per second. Nothing in the universe travels faster than that.

When light goes from air to water or into glass, it slows down and its path bends toward the surface of the water or window. When it comes back out, it bends the other way. That's why a straw in a glass of water looks crooked, like a "z." It's also why lenses in eyeglasses and telescopes are able to magnify images. Light shone through a teardrop-shaped lens is expanded when it comes out the other side.

White light is made of different colors. When you pass it through a prism, the colors become spread out because each one is bent by a slightly different amount.

ON THE FRONTIER

The invisibility possessed by the alien hunter in the *Predator* movies might not be so far-fetched. In recent years, scientists have learned to engineer special materials to bend light in ways that don't occur naturally. Researchers have proposed that a sphere of so-called metamaterial could render an object invisible by causing light to go faster than usual and slip around the sphere like air rushing over an airplane wing. In 2006, Duke University researchers made the first steps by showing that concentric disks of copper metamaterial could redirect microwaves around the middle. Let's hope we conquer invisibility before hostile aliens do.

COCKTAIL PARTY TIDBITS

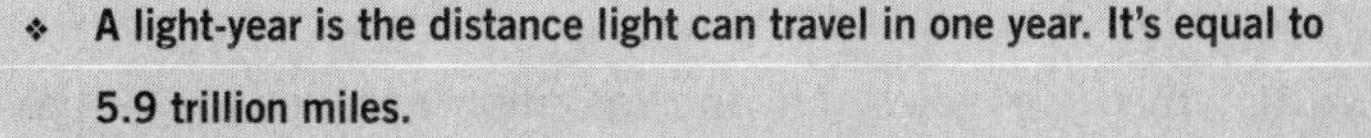

- **A light-year is the distance light can travel in one year. It's equal to 5.9 trillion miles.**
- **The speed of light is so fast it might as well be instantaneous to us, but out in space, it adds up. It takes eight minutes for light to reach us from the sun. The nearest star, Proxima Centauri, is 4.2 light-years away.**

THE ELECTROMAGNETIC SPECTRUM

THE BASICS

The universe is full of electromagnetic radiation, of which visible light is only a small part. The broader electromagnetic spectrum runs from radio and microwaves to x-rays and gamma rays.

As with any wave, light has a wavelength, which is the distance between crests and troughs in the rippling electromagnetic field. The different portions of the spectrum are defined by the wavelength of radiation. Light of shorter wavelengths carries more energy.

The visible spectrum runs from red at a wavelength of 750 nanometers to violet at 380 nanometers. Extending the spectrum beyond red, we come to infrared, microwaves, and then radio. Beyond violet is ultraviolet, x-rays, and gamma rays, which are called ionizing radiation because they have the energy to knock electrons from their nuclei.

If we looked at only the visible light coming from matter, we'd miss out on a lot. The only light that reaches us from some stars may be infrared or microwaves, not visible. Observing the universe at different wavelengths gives us a better understanding of what's going on.

ON THE FRONTIER

Remember those old x-ray specs advertised in the backs of comic books? See through people's clothing! Embarrass your friends! Modern technology has finally caught up with that dream, but the secret isn't x-rays: it's T-rays, or terahertz radiation, the portion of the electromagnetic spectrum between infrared and microwaves (wavelengths from one millimeter to one-tenth of a millimeter). Terahertz radiation passes harmlessly through clothing, but not metal. Some airports are testing terahertz security scanners to check passengers for guns, knives, and other objects. But isn't airport security bad enough without the TSA's electronic eyes strip-searching you?

COCKTAIL PARTY TIDBITS

- Microwaves are electromagnetic radiation with a wavelength between a few inches and a few feet.
- Scotch tape glows if you unroll it quickly in the dark. It also makes x-rays if peeled in a vacuum, as researchers discovered in 2008. They used the effect to perform an x-ray scan of a finger.
- Radio has the longest wavelength, up to miles or more. That's why radio telescopes are made of huge antenna dishes.

PHOTONS

THE BASICS

For centuries, researchers debated whether light was a wave or a particle. (These are the kinds of debates scientists have.) And in many ways, light does act like a wave. But just as seemingly smooth matter is composed of individual particles, light is actually made of particles called photons. Matter—specifically, electrons—can either gain energy by absorbing photons or lose it by spitting them out.

One key thing about photons is the amount of energy in one of them depends on the frequency of the electromagnetic wave or, in other words, the color of the light. Photons of higher frequency light carry more energy. That's why ultraviolet light damages skin and infrared light doesn't. The ultraviolet photons carry more wallop.

When light shines on some metals, it gives electrons a kick, creating an electric current. If light is purely a wave, a brighter light should give electrons more of a kick than a dimmer light. But experiments in the late 1800s found it was higher frequency light that mattered. Below a particular frequency, light doesn't kick out any electrons at all.

The other cool thing about photons is, unlike atoms, they can occupy the same spot at the same time. That's why laser light is so powerful. All the photons are concentrated in a small space.

ON THE FRONTIER

Lasers are a common fixture of modern technology, key ingredients in DVD players and fiber-optic communication lines that transmit broadband Internet connections. That's cool, but when will it be able to blow stuff up? Well, sooner than you might think. In August 2008, the defense contractor Boeing claimed it successfully tested a gunship-mounted laser cannon at Kirtland Air Force Base in New Mexico. Boeing says that by 2013, it plans to begin battlefield-strength tests of a truck-mounted system for blasting rockets, artillery shells, and mortar rounds.

COCKTAIL PARTY TIDBITS

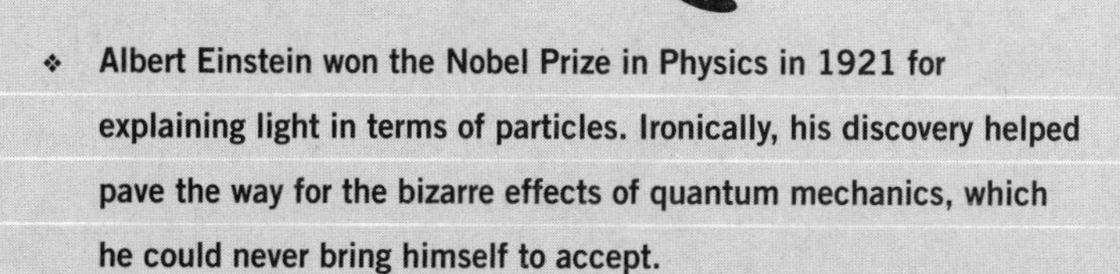

- **Albert Einstein won the Nobel Prize in Physics in 1921 for explaining light in terms of particles. Ironically, his discovery helped pave the way for the bizarre effects of quantum mechanics, which he could never bring himself to accept.**
- **Super lasers have peaceful purposes, too. At the National Ignition Facility in New Mexico, scientists are working on a set of 192 lasers designed to generate nuclear fusion for testing commercial laser-based fusion reactors.**

CHAPTER

THREE

QUANTUM WEIRDNESS

WHAT'S A QUANTUM?

THE BASICS

The theory of quantum mechanics is the bedrock of our understanding of atoms and subatomic particles, from the first moments of the Big Bang to today. It has a reputation for weirdness, but some of its key features are quite digestible. Quantum theory entails that the energy of matter must always come in bite-sized pieces of fixed size. The word *quantum* means an indivisible unit. A particle may contain, say, three or four units or "quanta" of energy, but never an in-between value, such as 3.5.

We don't see these quantum chunks of energy for the same reason we don't see the pixels in a printed photograph. Compared to the energies we're used to, they are small. They turned up when researchers began to study the atom in detail. An electron's distance from the nucleus depends on the amount of energy it contains. But if an electron can only have particular energies, it can't climb to a higher orbital like a rising airplane. This is why electrons do not in fact fly around the nucleus the way planets do.

The electron is more like an elevator. A pair of orbitals is separated by a particular amount of energy, like the distance between floors. To jump up to the higher orbital, an electron has to absorb a photon containing exactly that amount of energy: no more, no less. Once it's there, it can only drop down to the lower orbital by emitting exactly that energy, again as light.

ON THE FRONTIER

We can't watch an atom the same way we would a planet, because shining light on it disturbs its electrons. But scientists can fire off extremely short pulses of laser light to "ping" an atom at a precise instant to see how it responds. The effect is something like a strobe light. Researchers at Germany's Max Planck Institute for Quantum Optics have used pulses lasting only a few hundred attoseconds (an attosecond is a billionth of a billionth of a second) to study how atoms and molecules become ions. Future experiments may allow them to watch orbitals changing shape during chemical reactions.

COCKTAIL PARTY TIDBITS

- **If an electron could emit energy in any amount, it would spiral into the nucleus in less than a trillionth of a second, firing off deadly gamma rays in the process.**
- **The bite-sized nature of quantum mechanics is expressed in terms of a very small number called Planck's constant, named for German physicist, Max Planck.**

WAVE-PARTICLE DUALITY

THE BASICS

We've seen that light is both a wave and a particle. Could the same be true for subatomic particles? You bet it is. One way of thinking about electrons in an atom is as a guitar string wrapped in a circle. If you plucked the string, it could only vibrate at particular frequencies, the same way a guitar string of a given length can only make certain notes.

The wave nature of particles is by no means obvious. The clearest way to see particles acting as a wave is to shoot a bunch of them through a set of narrow openings. If a wave is forced through openings smaller than its wavelength, it will form a set of crisscrossing ripples, like two stones dropped side by side in a pond. This is how scientists first proved that light acts like a wave. When shone through narrow openings, it forms a pattern of bright and dark bands called interference fringes.

We don't notice the waviness of electrons and other subatomic particles, because their wavelengths are very short. But using sophisticated techniques, researchers have observed the same diffraction pattern for electrons that they observed for photons. The same goes for neutrons and even some molecules.

ON THE FRONTIER

Maybe you've heard of an electron microscope. If you've ever seen an image of bacteria or viruses that looked like a frozen landscape in black and white, that's an electron microscope image. In a light microscope, light waves reflecting off a surface reveal any little surface features that are at least as big across as the wavelength of the light. An electron microscope works the same way, but it's more powerful, because electrons have a shorter wavelength. Light microscopes can magnify only about 2,000 times. An electron microscope can magnify images up to two million times.

COCKTAIL PARTY TIDBITS

- To prepare a living sample for electron microscopy, scientists have to prepare the sample in special ways, such as coating the surface of a bee in gold or flash-freezing a sample in liquid nitrogen and then splitting off a piece.
- Every year, imaging-technology companies sponsor competitions for the best microscopic images of living things. The winning entries often use electron microscopes. Check out Nikon's "Small World" contest online.

WAVES OF PROBABILITY

THE BASICS

Here's where we try to blow your mind a little. In the experiment that produces interference fringes, let's say we dimmed our beam of light so that only one photon went through the slits at a time. We'd see each particle land in one spot at random (where the particle detector absorbed one quantum of energy). As more and more photons went through, they would gradually build up the diffraction pattern.

In the everyday world, if you flip a coin, it lands randomly on heads or tails. But if you knew precisely the position of the coin and the forces on it when you flicked it, as well as the exact position of the surface it landed on, you could in principle use a powerful supercomputer to predict how it would land.

But in the quantum world, there's nothing you can use to predict where a particle will end up after it passes through the tiny slits. It's truly random. All we can say is a particle is described by something called a "wave function" that tells you how likely it is that a particle will land here or there. The same goes for other events in the quantum world. You can't predict which photon will pass through a window and which will be reflected, or when an electron will emit a photon.

ON THE FRONTIER

Before quantum mechanics, researchers thought that if they could set up an experiment in exactly the same way twice, they would always get the same outcome. Quantum mechanics was a major break from that ideal. Einstein thought the theory was missing some key ingredient that would explain the randomness in more familiar terms, and there are researchers today who still agree with him. But nobody can argue with the success of quantum mechanics in predicting what happens in the subatomic world.

COCKTAIL PARTY TIDBITS

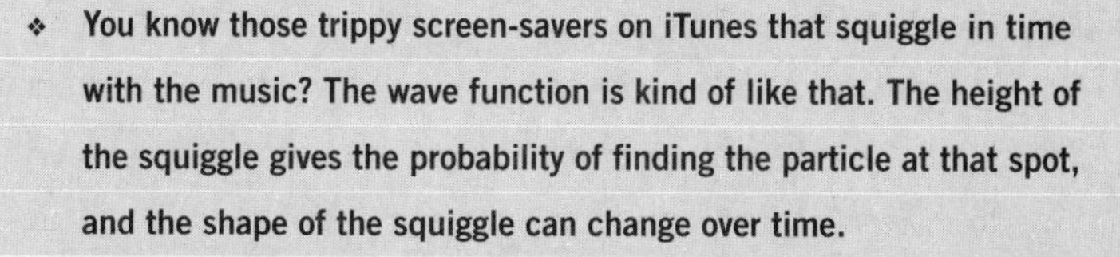

- **You know those trippy screen-savers on iTunes that squiggle in time with the music? The wave function is kind of like that. The height of the squiggle gives the probability of finding the particle at that spot, and the shape of the squiggle can change over time.**
- **Einstein used to argue with his Danish colleague Niels Bohr, a co-discoverer of quantum mechanics. Einstein kept telling Bohr, "God doesn't play dice," until one day Bohr retorted, "Quit telling God what to do."**
- **Most scientists don't worry too much about what quantum mechanics means, at least not during work hours. They adhere to the motto, "Shut up and calculate."**

SUPERPOSITION

THE BASICS

People often say that in the quantum world, a particle can be in two places at once, a state called a superposition. The diffraction experiment with photons is a convenient way of thinking about superposition, because we can limit the number of gratings to just two, one left and one right. The wave function goes through both openings. We say the photon is in a superposition of both states—left and right—until we measure its position and it picks a path at random.

The weird thing is, if you try to peek and see which path the photon "really" took by setting up your experiment to distinguish between the two paths, the superposition goes away and the interference pattern disappears. Scientists sometimes say the act of measurement "collapses" the wave function. Basically, if you look for a particle, you will find it in one spot or another. If you don't look for it, it will take every possible path it can, all of which will contribute to the probability that it ends up in one place or another.

Human intervention doesn't really have anything to do with it, though. If something is in a superposition, it just means it's briefly isolated from its surroundings. As soon as another particle bumps it, the superposition goes away.

ON THE FRONTIER

One thing you might do with superpositions is make a really powerful computer. Normal computers run on 1s and 0s. But in something called a quantum computer, atoms would be put in superpositions of two states, in effect acting as 1 and 0 at the same time. By doing that to a bunch of atoms and making them work together, a quantum computer might solve problems that are really hard today, such as cracking secret codes or simulating how materials work.

COCKTAIL PARTY TIDBITS

- A Bose–Einstein condensate is a cloud of gas chilled to near absolute zero so that all the gas atoms share the same superposition, making them effectively one big atom.
- Scientists used to ponder whether an object as big as a cat could be in a superposition—Schrodinger's infamous cat, named for German physicist Erwin Schrodinger. But a cat is full of particles bumping together, so they can't all be in one big superposition.
- But if a wave function collapses in the woods and there's nobody there to measure it. . . . Oh, never mind.

THE UNCERTAINTY PRINCIPLE

THE BASICS

Quantum mechanics severely limits what we can know about a particle. Whenever you think you have a particle's identity pinned down, it finds a way to wiggle out.

Say you were to confine an electron in a small space, which would allow you to know its position very precisely. If you measured its momentum repeatedly, it would be all over the map. It's as if particles are claustrophobic. Confined to small areas, they start climbing the walls like mad trying to get out. The more you hem in a particle, the more violently it will jump around.

This is the meaning of the famous Heisenberg uncertainty principle, named for German physicist Werner Heisenberg, which says no experiment can reveal both the position and the momentum of a particle.

So does a particle really have both a position and a momentum, but we can only measure one at a time? Scientists don't think so. When you measure one property, the other one actually fuzzes out, and the more precisely you measure the first thing, the more fuzzed out the second one gets. Particles are like those absent-minded eggheads who can only keep one thing on their mind at a time.

ON THE FRONTIER

The uncertainty principle applies in the same way to energy and time. Researchers can chop laser light into short-duration bursts. But when they do, they lose control over the amount of energy in each pulse. They start off with laser light of a very precise energy, which corresponds to a particular frequency or color. The color of the pulse is smeared out. It's as if the universe can't keep track of energy when things are happening fast.

COCKTAIL PARTY TIDBITS

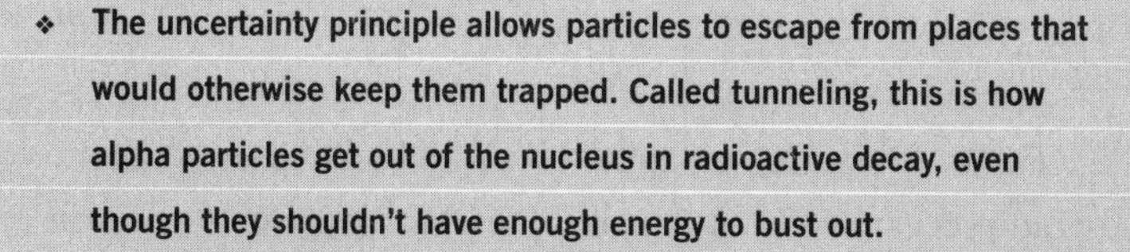

- **The uncertainty principle allows particles to escape from places that would otherwise keep them trapped. Called tunneling, this is how alpha particles get out of the nucleus in radioactive decay, even though they shouldn't have enough energy to bust out.**
- **In the *Star Trek* universe, transporters work because of technology called a Heisenberg compensator, which removes quantum uncertainty about the atoms being teleported.**

ENTANGLEMENT

THE BASICS

The randomness of quantum mechanics is weird enough. But the weirdness doesn't stop there. According to quantum rules, the behavior of a particle on one side of the universe can determine that of a second particle on the other side of the universe, instantaneously, without any signal traveling between the two.

This quantum link is called entanglement, and Einstein had a hard time with it. He called it "spooky action at a distance" and thought it was a sign that quantum mechanics was not the last word on the universe. By now, researchers have come to accept it, even if they don't understand it.

Particles become entangled when they are born from the same process, such as in special crystals that split a photon into two lower energy photons. If researchers make lots of photon pairs and measure each polarization of each photon (how it's oriented in space) they'll get random results from one pair to the next.

But they will find that the measurements for each pair are always related in a particular way. If one photon in the pair is polarized vertically, the other one will be polarized horizontally. It's like looking at the dots in a stereogram image. By themselves, the dots look random. But look at two sets of dots, one in each eye, and an image emerges.

ON THE FRONTIER

Scientists have used entanglement to create a kind of teleportation. They can't make matter literally dissolve and reappear somewhere else. But they can do the next best thing. Researchers have used entanglement to "teleport," or transfer, the quantum state of a beam of light onto a cloud of atoms, the kind of thing that might be useful in a quantum computer. Just don't count on getting beamed up by Scotty anytime soon. Teleportation only works for things that are in nice, clean quantum states, not for messy living things.

COCKTAIL PARTY TIDBITS

- **You can use entanglement to send secret codes. If an eavesdropper tries to intercept it, she will disturb the entanglement and give herself away.**
- **In 2007, researchers transmitted entangled photons 89.5 miles through the air between the Canary Islands of La Palma and Tenerife, off the coast of Morocco. Next up: bouncing entangled light thousands of miles from Earth to the International Space Station and back, perhaps as soon as 2014.**

QUANTUM UNREALITY

THE BASICS

You might wonder if the strangeness of quantum mechanics is for real or if we're missing some other explanation that's more in tune with our everyday understanding of the world. Quantum theory has us convinced that particles have no true properties such as position or momentum, and that entanglement is an instantaneous link.

But maybe a particle has real properties that are hidden somehow from our experiments. And so, when a pair of entangled particles is born, maybe each member of the pair is imprinted with instructions telling it how to respond when measured in each of a few different ways. The two particles receive complementary instructions, so there's no need for any instantaneous link between them. Each particle hits a detector and acts on the instructions it received, and that's that. This idea is called local realism.

The Irish physicist John Bell devised an experiment that would distinguish the two possibilities. Known as the Bell test, it involves sending each member of a pair of entangled particles, typically photons, into a separate detector. In the case of photons, the detectors measure the spatial orientation, or polarization, of each photon's electric field. The polarizations of the pair will either match or not. If local realism is correct, the polarizations should not match more than a certain percentage of the time. But they do, according to repeated experiments.

ON THE FRONTIER

Entanglement seems to violate the spirit, if not the letter, of Einstein's special theory of relativity, which says nothing travels faster than light. But to confirm that entanglement has taken place, researchers have to compare notes from each member of the pair of entangled particles. And that at least can't happen faster than light.

For those who think the awkwardness of quantum mechanics is here to stay, another approach would be to try to understand how entangled particles view space and time. Maybe entanglement is pointing to a deeper theory of the universe.

COCKTAIL PARTY TIDBITS

- **In a 2008 experiment, researchers found that if entanglement is not truly instantaneous, it operates at speeds at least 10,000 times faster than light.**
- **A more elegant version of Bell's test is to create three photons that are all entangled with one another. In this version of the experiment, it takes only a few measurements to confirm that quantum mechanics is right.**

CHAPTER

FOUR

MOTION, SPACE, AND TIME

MASS AND INERTIA

THE BASICS

In school, everybody learns Newton's laws of motion, named for the great seventeenth-century English scientist Sir Isaac Newton. They describe the behavior of moving objects in the everyday world, from billiard balls and bullets to Olympic sprinter Usain Bolt.

Newton's first law famously states that an object in motion stays in motion, or if it's at rest it stays at rest, unless acted on by an outside force. On Earth, when you kick a ball, it knocks into molecules in the air and ground, leading to the forces of air resistance and friction, respectively, which eventually bring it to rest.

But if you kicked a ball in space, it would keep going at the same speed forever, or at least until it hit an asteroid or something. It needs absolutely no help to keep moving. We say the ball has inertia. This concept is known as Newton's first law.

Inertia is proportional to mass. The more mass something has, the more of a push it needs to get going. Weight is not the same thing as mass. Weight is a product of Earth's gravity. Objects in space may not have weight, but they still have inertia. That's why you should duck if you're ever an astronaut and you see a bowling ball coming at you. Never mind how it got there.

ON THE FRONTIER

After hundreds of years, scientists are finally pretty sure they know the origin of inertia and mass. Brace yourself. They think there's a field, sort of like an electric or magnetic field, that fills all of space and acts like Jell-O through which subatomic particles have to plow. It's called the Higgs field, and some particles have an easier time getting through it than others, sort of as if they're wearing snow shoes, or maybe Jell-O shoes. So why do some particles feel this Jell-O more than others? Oy, don't ask. Nobody is sure.

COCKTAIL PARTY TIDBITS

- **There is a funny video on YouTube that shows inertia in action: a semi-truck is pulled up to a Target loading dock, full of shopping carts. It pulls away before workers close up the truck's back door, so dozens and dozens of shopping carts go falling out.**
- **The famous Italian scientist Galileo Galilei first deduced Newton's first law by rolling a ball down a series of smooth ramps.**
- **Mass is measured in units called kilograms. One kilogram is defined as the mass of a highly polished platinum–iridium ingot, seven copies of which are stored at France's International Bureau of Weights and Measures.**

FORCES AND ACCELERATION

THE BASICS

A force is anything that causes acceleration, a change in speed or direction. Newton's second law says for a given object, such as a kickball, it takes more force to accelerate it to higher speed. So to make a kickball go farther, you have to give it a larger or more sustained force—in effect, a harder kick. Similarly, if you push a marble and a bowling ball with the same force, the marble will pick up more speed.

On Earth, gravity pulls everything down at the same rate. When you jump into the air, you will fall back down at a rate of 32 feet per second per second. So if you jump out of an airplane, for every second you fall, your speed will increase by 32 feet per second.

Newton's third law says that if a force going one way encounters resistance, there must be a second equal force acting in the other direction. Think about driving a car. When you hit the gas, your body's inertia resists the change in motion. You slide backward into the seat, which presses back with equal force. Otherwise you or it would be in motion. That's Newton's third law.

ON THE FRONTIER

You never know when you'll need an instrument to measure acceleration, called an accelerometer. They are commonly used in rockets and space telescopes to maintain the proper trajectory and orientation. But they are showing up more often in everyday gadgets. They are used in laptops to sense falls; in iPhones to know when you've flipped it around; and in the Nintendo Wii controller to punch, slash, and serve. Some teachers in Italy have started bringing the Wii controller into the classroom to conduct simple experiments such as measuring the acceleration of gravity. We hope they work a little better for science experiments than for picking up those subtle flicks of the wrist in bowling.

COCKTAIL PARTY TIDBITS

- **Newton's third law tells us that the force holding molecules together is much stronger than gravity, otherwise we would go crashing through the floor.**
- **Accelerations are measured in multiples of gravitational or g-forces. When you sneeze, you are pulling almost four g.**
- **The human body can withstand 16 g for perhaps up to a minute. Racecar driver Jeff Gordon supposedly did 64 g when he crashed his car in the 2006 Pennsylvania 500.**

MOMENTUM

THE BASICS

Newton's laws of motion don't apply to subatomic particles and for objects moving near the speed of light. The theories of quantum mechanics and special relativity had to be formulated to describe motion in those situations. These broader theories are not concerned with forces, but with something called momentum.

Once an object is moving, we say it has acquired momentum. It's basically inertia pointed in the direction the thing is moving. The more mass in an object and the faster it's going, the more momentum it has. And the longer a force acts on something, the more momentum it picks up. That's why you should always follow through in baseball, to prolong your bat's contact with the ball until the last moment.

Like energy, momentum has to come from somewhere. It is always conserved. That's why guns recoil when they are fired. The gun and bullet start out with zero total momentum. They can still have zero total momentum when the bullet is fired, if the momentum of the bullet in one direction cancels the momentum of the gun in the other direction.

There's also angular momentum, which you've seen at work in figure skating routines. Conservation of angular momentum is why a skater spins faster when he pulls in his arms. A fast-spinning, compact object has the same angular momentum as a slower spinning, extended object of the same mass.

ON THE FRONTIER

Astronomers see conservation of angular momentum at work all the time in outer space. It seems to be the driver behind jets of plasma that spray from young stars and dense stellar objects such as pulsars and black holes. These dense customers pull disks of rotating gas and dust around them. Often the inner part of the disk will begin to lose angular momentum as it falls onto the star or other object. The angular momentum has to go somewhere. A convenient outlet is a pair of high-speed jets shooting from both ends of the star, which creates a lot of momentum without much mass.

COCKTAIL PARTY TIDBITS

- **Would you believe that light has momentum? Researchers are working on so-called solar sails that would harness this momentum to propel spacecraft using sunlight or laser beams.**
- **Somewhere else you've seen conservation of momentum is in those video games where you knock a little ball back and forth using a paddle on either side of the screen. If your paddle is still when you hit the ball, its direction doesn't change.**

GRAVITY

THE BASICS

Gravity is the force that keeps us planted on the ground, maintains Earth's roundness, and makes the planets swing around the sun. In 1687, Isaac Newton, one of the premier eggheads of all time, published his law of universal gravitation, which says the gravitational force between two objects depends on three things: their respective masses, the distance between them, and the intrinsic strength of gravity.

Newton's law explained the orbits of the planets as well as why a bowling ball and a golf ball fall at the same rate, despite their different masses. The force between Earth and the bowling ball is more than the force between Earth and the golf ball. So they both fall at the same rate.

Newton's law says all masses attract one another, and as long as the objects are far enough apart, the size of the force depends on the distance between masses squared. Double the distance between the masses and you quarter the force. On Earth, the distance between masses doesn't matter much because everything is so small compared to Earth.

One thing that's not small compared to Earth is the moon. The gravitational attraction between Earth and the moon is responsible for the tides. The moon pulls the water facing it closer to it, or in other words, high tide tends to follow the moon.

ON THE FRONTIER

Gravity is responsible for the structure and shape of the universe as a whole. It must be a very powerful force, right? Yes and no. On the scale of the atom it's actually quite feeble. Until the 1990s, scientists were unable to measure the force of gravity between two objects separated by a few millimeters or less. Gravity is so weak that experiments just weren't sensitive enough. (Ongoing tests are measuring the strength of gravity at even shorter distances.)

But unlike the force between charged particles, gravity is cumulative, because all objects are gravitationally attracted to one another. After a while, even a cloud of gas floating in space will collapse under its own gravity, forming stars and planets, galaxies, and even clusters of galaxies.

COCKTAIL PARTY TIDBITS

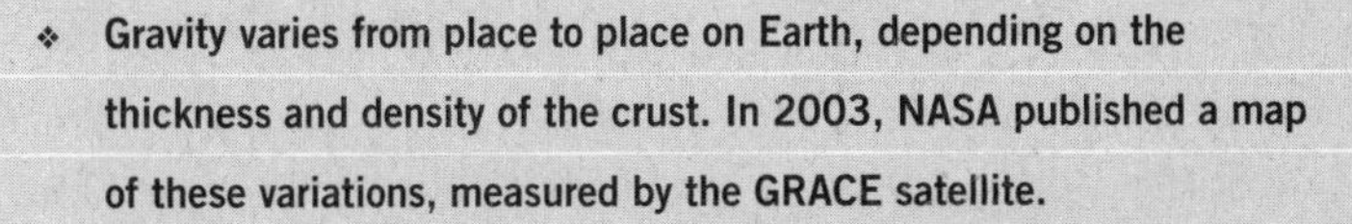

- **Gravity varies from place to place on Earth, depending on the thickness and density of the crust. In 2003, NASA published a map of these variations, measured by the GRACE satellite.**
- **The gravity of a spheroid object such as the Earth decreases as you dig beneath the surface. If you could travel to Earth's core, you'd find that your weight would gradually decrease to zero at the very center of the planet.**

ORBITAL MOTION

THE BASICS

In developing his law of gravity, Newton's great insight was to see that the force pulling an apple to Earth is the same force holding planets in their orbits.

The sun is like a child swinging a bucket of water. The bucket is moving in an arc, but the force on the bucket is not pushing the bucket sideways; it's pulling it toward the child. That's why the water is pressed against the bottom of the bucket; it's Newton's old equal-but-opposite-forces law. The inward swinging force is matched by an equal but opposite outward force.

The planets work the same way. Gravity pulls them toward the sun. To see how that works, let's imagine that the child is now the Norse god Thor, who can whip his trusty hammer Mjolnir a good ways by spinning and letting it go. The harder Thor throws the hammer, the farther it will go before gravity brings it back to Earth.

But because Earth is curved, if Thor chucks Mjolnir at just the right speed, it will fall toward Earth at the same rate Earth drops away ahead of it. It has achieved orbit. The planets are doing the same thing. Each one is falling toward the sun at a speed that keeps it in orbit.

ON THE FRONTIER

With his insight, Newton was able to explain the three laws of planetary motion developed by astronomer Johannes Kepler in 1605. Kepler had discovered that each planet traces an ellipse, moving faster as it gets closer to the sun, with an orbital period that depends on the size of the ellipse.

Kepler's laws guide astronomers today in their studies of distant stars and planets. Planets around other stars are so distant we can't see them directly. But researchers have tricks to figure out how long those planets take to orbit their stars, which tells them, by Kepler's third law, the distance between planet and star.

COCKTAIL PARTY TIDBITS

- **Before Newton and Galileo, everybody thought the planets moved because angels were pushing them.**
- **The speed at which Thor would have to throw his hammer to put it in orbit depends only on Earth's mass. Called Earth's orbital velocity, it is seven miles per second. To completely escape the solar system from Earth (i.e., to overcome the sun's gravity) Thor would have to heave Mjolnir at a speed of 26 miles per second.**
- **Researchers have proposed building a space elevator, a cord extending from the ground into space, held taut by a counterweight orbiting more than 22,000 miles up.**

SPECIAL RELATIVITY

THE BASICS

We need the theory of special relativity to understand what happens when objects start to travel at speeds close to that of light. Say you are on the Battlestar *Galactica* trying to outrun nuclear-tipped missiles that a Cylon basestar just fired at you. Your hyperspace drive is damaged, so you hit the thrusters. The speed of the incoming nukes relative to *Galactica* has slowed, buying you a few more precious seconds as engineers work to reboot the hyperdrive.

You wouldn't have been so lucky if the Cylons had finished building their new laser weapon, capable of melting a hole straight through even a battlestar-class ship. Albert Einstein didn't know Admiral Adama from Adam, but the central insight of his theory of special relativity would apply all the same. After pondering Maxwell's equations for electromagnetism, Einstein realized that the speed of light must always be the same, no matter how fast you are moving with respect to the light source.

Maxwell's equations don't care how fast a light source is moving. The speed of light is the speed of light. For the *Galactica* crew, that means they would find it much harder to buy time after the Cylons fired their death ray, because it would keep coming for Galactica at the speed of light.

ON THE FRONTIER

Say the Cylons had their laser working and *Galactica* had a good head start. If the ship could somehow build up a lot of speed between it and the Cylons before the laser hit, say 10 to 20 percent the speed of light, the laser still wouldn't slow down but it would lose some of its energy. When you are moving away rapidly from a light source, the light waves actually get stretched in something called the relativistic Doppler effect. The regular Doppler effect is what makes an ambulance siren rise in pitch as it approaches and fall as it recedes. Red light has a bigger wavelength than blue light, so we say the light is redshifted.

COCKTAIL PARTY TIDBITS

- **In Einstein's day, some researchers thought an invisible "æther" must transmit light waves in the same way that water transmits ocean waves. As Earth traveled through the aether, the speed of light would speed up and slow down, like an ocean ship moving with or against the waves. But experiments failed to detect the predicted change.**
- **Relativity is a misleading name. The basic idea is that the laws of nature are the same no matter how fast you are moving relative to anything else in the universe.**

TIME DILATION

THE BASICS

Battlestar Galactica is pretty geeky, but we can do better. Everything you really need to know about the surprising consequences of special relativity can be learned from the Disney movie, *Flight of the Navigator*. In this 1986 flick ("classic" might be too strong a word), an alien spaceship abducts David, a twelve-year-old boy, and flies off into space with him at high speed. When David returns, twelve years have passed on Earth, but he has aged only a few hours.

That in a nutshell is the phenomenon of time dilation, the slowing of time at high speeds. Special relativity dictates that when something moves very fast relative to you, time will pass slower for it than you, be the moving object a particle, a stopwatch, or a galaxy. If David's family could have observed him on his journey through a telescope, he would have appeared to them to be moving in slow motion. The greater the speed is, the more time slows.

Hand in hand with time dilation comes a distortion of space called length contraction. If David's family could have watched the spaceship buzz past Earth, it would have appeared flattened like a pancake along the direction it was traveling. David himself would have looked like a stick figure. So not only would near–light speed travel keep you young, you'd be skinny, too.

ON THE FRONTIER

Special relativity seems weird to us because we don't regularly go flying around in spaceships at near the speed of light. But if trips near light speed were commonplace, we'd get used to traveling forward in time. This must have been the concern of celebrity heiress Paris Hilton, who last year booked a flight on the Virgin Enterprise Rocket, a space-tourism venture. "With the whole light-years thing," she told reporters, "what if I come back 10,000 years later, and everyone I know is dead?" For that to happen, Virgin would have to arrange for its rocket to travel at 99.9999999 percent the speed of light for a ten-hour flight.

COCKTAIL PARTY TIDBITS

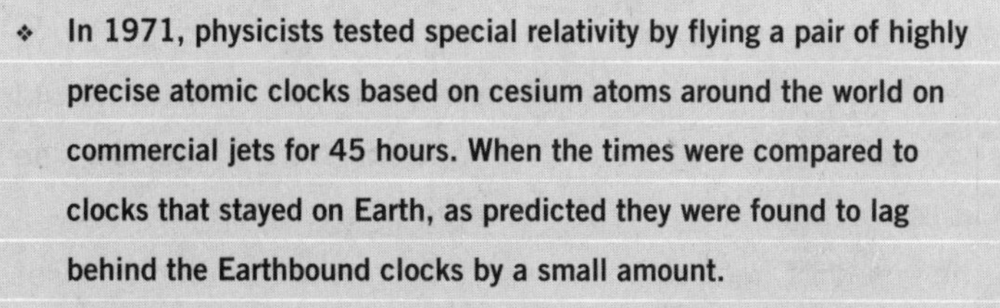

- **In 1971, physicists tested special relativity by flying a pair of highly precise atomic clocks based on cesium atoms around the world on commercial jets for 45 hours. When the times were compared to clocks that stayed on Earth, as predicted they were found to lag behind the Earthbound clocks by a small amount.**
- **Unstable particles called muons are constantly flying through Earth's atmosphere; see the section on cosmic rays. They would normally break down after 2.2 microseconds, but moving at 99.8 percent light speed, they last 35 microseconds by Earth's clock, or long enough to travel 6,000 miles.**

MASS-ENERGY AND THE LIGHT SPEED LIMIT

THE BASICS

According to special relativity, anything that has mass will never reach light speed, let alone exceed it. To see why, it's time to unpack Einstein's famous equation, $E=mc^2$, which we've thrown around as if it should be perfectly obvious what it means for mass and energy to be interchangeable.

Einstein's equation means that whenever an object gains or loses mass, it also gains or loses a relatively tiny amount of energy, and vice versa. If you push a boulder up a hill, it gains gravitational potential energy and therefore a little bit of mass. When the sun radiates energy in the form of light, it loses some mass. Get the idea?

The equation $E=mc^2$ refers only to the energy locked up in an object at rest. We call the m in the equation an object's rest mass. It takes a big change in energy to make a noticeable dent in the rest mass. The equation states that a change in rest mass is equal to the change in energy divided by c^2, the speed of light squared, which is a large number.

But the amount of energy you measure in something changes if it is moving relative to you. Special relativity dictates that as a spaceship approaches the speed of light, its energy, as measured by a stationary observer, approaches infinity. To reach light speed the energy would have to hit infinity, which can't happen.

ON THE FRONTIER

So how close to the speed of light can we get? It depends on the mass involved. Particle accelerators use powerful magnets to accelerate protons and electrons in excess of 99.99 percent light speed. We might be able to reach 10 to 20 percent light speed in a spaceship powered by nuclear explosions. A solar sail powered by momentum from a laser beam could go much faster because it wouldn't need fuel, but it couldn't carry passengers either. Our best bet for interstellar travel might be "generation ships," big freighters that would house miniature civilizations until they reached their destinations.

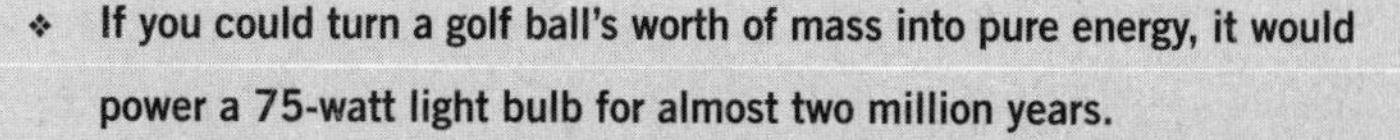

COCKTAIL PARTY TIDBITS

- **If you could turn a golf ball's worth of mass into pure energy, it would power a 75-watt light bulb for almost two million years.**
- **Sometimes on *Star Trek* they would talk about tachyons ("tacky-ons"). These are hypothetical particles that move faster than light but can never dip below light speed, which is not prohibited. But it's not clear tachyons could exist in the real world.**

SPACETIME

THE BASICS

Because of time dilation and length contraction, a pair of events that seem simultaneous from one perspective will seem to happen at different times from another perspective. Special relativity tells us how to translate between those perspectives, by locating them in spacetime, which is the same for everybody. We mark an event in spacetime—the explosion of a star, a birthday party—by its location and when it happened, as judged from somebody's (anybody's) perspective.

Spacetime is like a loaf of raisin bread. Being at rest is like cutting the bread into slices in the usual way. The length of the loaf is like time, and the width and height are like distance in two directions. An event is a raisin. Each slice is one second "thick" and contains a particular number of raisins, which correspond to simultaneous events.

If somebody is moving relative to you, it's as if they turned the loaf before slicing it up. Raisins that were in a single one of your slices may be spread out between two, three, or more of their slices. But thanks to the equations of special relativity, no matter how the slices are cut, when they are all put together, the raisins are in the same spots.

ON THE FRONTIER

Although space and time individually are relative, spacetime is not. It's the same for everybody across the universe. Distant galaxies are moving away from us at high speed, so you might think aliens in that galaxy would disagree about the time elapsed since the Big Bang, but as it turns out, other galaxies aren't moving away from us through spacetime; they're being carried along as spacetime expands. We have more to say about this later, but for now it just means that everybody agrees on the age of the universe: 13.7 billion years.

COCKTAIL PARTY TIDBITS

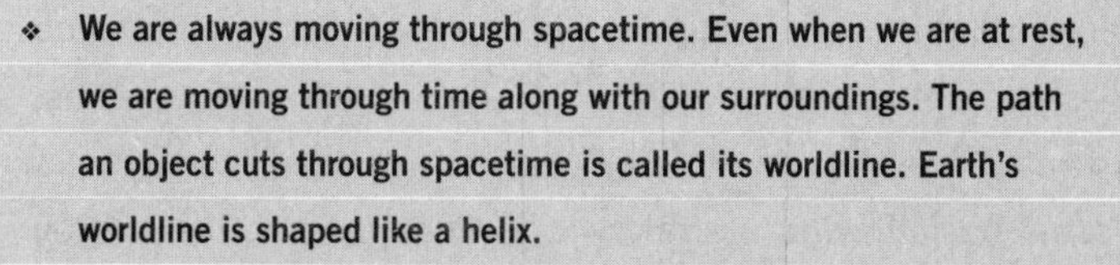

- **We are always moving through spacetime. Even when we are at rest, we are moving through time along with our surroundings. The path an object cuts through spacetime is called its worldline. Earth's worldline is shaped like a helix.**
- **There could be extra dimensions of space that are "curled up" so small we can't see them. They would be like little pockets you can only enter if you shrink down to their size.**

GENERAL RELATIVITY

THE BASICS

The general theory of relativity is Einstein's theory of gravity. It says gravity is unlike the other forces in that it's not dragging objects kicking and screaming against the grain of space and time. Gravity works with spacetime. In fact, it is spacetime. Einstein's realization was that massive objects such as the sun actually bend spacetime, like a bowling ball on a trampoline. Smaller objects such as planets follow this curvature naturally, the way a tennis ball would if you rolled it past the bowling ball.

Einstein's realization came from thinking about the weightlessness of freefall. A skydiver doesn't feel any force of gravity during freefall. If she were blindfolded, she might as well be floating in space. It takes a force—in the form of the ground—to stop this freefall. Einstein decided the skydiver must be following a straight line through spacetime. Now, imagine hordes of skydivers falling at different points around the globe. Their paths all intersect at Earth's core. Our planet must cause "straight" lines to converge at its center.

According to general relativity, concentrated mass slows the time around it and distorts space, too. If mass becomes concentrated enough, like in a star such as the sun, it can even tug passing light like a car trying to drive straight along a steep bank.

ON THE FRONTIER

When two massive objects such as a pair of black holes collide, general relativity predicts that spacetime will ripple like a pond. These waves would spread at the speed of light and have the effect of first stretching spacetime in one direction and then compressing it, like an accordion. To detect gravitational waves, scientists have built a big experiment called LIGO, consisting of two 2.5-mile-long vacuum chambers spaced more than 1,800 miles apart. Any gravitational waves passing by would shift the laser light shone down the 2.5-mile pipes.

COCKTAIL PARTY TIDBITS

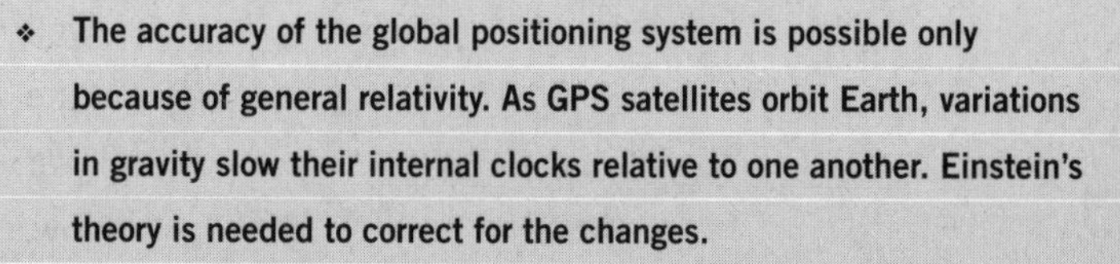

- **The accuracy of the global positioning system is possible only because of general relativity. As GPS satellites orbit Earth, variations in gravity slow their internal clocks relative to one another. Einstein's theory is needed to correct for the changes.**
- **Einstein became famous after astronomers made a historic test of general relativity during a total eclipse in 1919. They detected changes in the positions of stars near the sun as it passed through the sky, a sign that the sun's gravity was bending light around it.**

CHAPTER

FIVE

THE SOLAR SYSTEM

THE SUN

THE BASICS

Everything in the solar system, from the planets to the lowliest meteor, orbits the sun, the big ball of glowing gas visible daily between sunrise and sunset. Measuring 109 Earths across, the sun accounts for 99.8 percent of the mass in our entire solar system. Gravity squeezes the sun's core to such great pressure that hydrogen nuclei fuse into helium, releasing energy that cooks the core to a temperature of 27 million degrees Fahrenheit.

By the time that radiation works its way to the surface of the sun, it still has enough energy to heat the atoms there to a temperature of 10,300 degrees, causing them to radiate light that illuminates and warms the planets, including Earth. Studying that light tells us the atoms there are 75 percent hydrogen, 24 percent helium, and about 1 percent trace elements, including iron, nickel, oxygen, silicon, sulfur, magnesium, carbon, calcium, and chromium.

The sun is a fusion machine. Every second, it crushes 700 million tons of hydrogen into 695 million tons of helium. Sounds impressive, but that's only a tiny fraction of Earth's mass. In its 4.6-billion-year life, the sun has burned through only 7.8 percent of its hydrogen. Researchers expect it to keep cooking for another five to six billion years.

ON THE FRONTIER

Surrounding the sun for millions of miles is the corona, a halo of charged particles swept out by the sun's powerful magnetic field. The temperature of the corona is some 3.6 million degrees, hundreds of times hotter than the surface. Something must be pumping energy into the corona. Researchers suspect it's some mixture of recombination (the snapping and rejoining of magnetic field lines) and waves in the magnetic field that compress the plasma, sort of like snapping a towel. NASA has proposed the Solar Probe Plus mission to fly a spacecraft within a few million miles of the sun to check things out. Still in the design phase, the mission could launch as early as 2015.

COCKTAIL PARTY TIDBITS

- **It can take a million years for photons inside the sun to emerge to the surface. Getting from there to Earth takes a mere 8.3 minutes.**
- **The sun has a vast atmosphere of charged particles—the solar wind—that extends beyond the orbit of Pluto. It finally peters out in a region called the heliopause, where the wind has lost the energy to rebuff the material between stars.**
- **Don't call the sun average. It ranks among the top 10 percent in mass in our galaxy; of the 50 nearest star systems, ours is the fourth brightest.**

THE PLANETS

THE BASICS

The eight planets of the solar system formed some 4.6 billion years ago from a "protoplanetary disk" of gas and dust swirling around the sun. As dust is wont to do, it clumped into dust bunnies, snowballing into great boulders over millions of years and fusing into numerous moon-size protoplanets, which collided like pinballs until the eight planets formed.

Grains of rock and metal gravitated toward the center of the disk, creating four rocky or "terrestrial" planets: Mercury, Venus, Earth, and Mars. Each one has a dense core of iron surrounded by a mantle rich in silicon dioxide, the same stuff of which glass, quartz, and concrete are made.

Radiation from the sun blew gases farther out to mingle with more distant bits of icy rock, forming the four gas-giant planets: Jupiter, Saturn, Uranus, and Neptune. Also called Jovian planets after Jupiter, their helium and hydrogen atmospheres gradually liquefy, so these planets lack distinct surfaces. They have rocky cores, which aren't so much miniplanets as clusters of iron and nickel mixed with lighter elements.

The leftovers of planet formation became asteroids, comets, and the icy bodies of the Kuiper Belt, including Pluto and other dwarf planets, which achieved roundness but never grew big enough to sweep up everything in their path.

ON THE FRONTIER

The concept of a dwarf planet is still new and is probably subject to revision. The International Astronomical Union demoted Pluto to dwarf status in 2006 after the discovery of Eris, an even larger object in the Kuiper Belt. The IAU now recognizes five dwarf planets: Pluto, Eris, Haumea, Makemake (all in the Kuiper Belt beyond Neptune), and Ceres (an asteroid). There could be many more, but even this group is provisional. Astronomers have studied only Ceres and Pluto closely enough to be sure of their status.

COCKTAIL PARTY TIDBITS

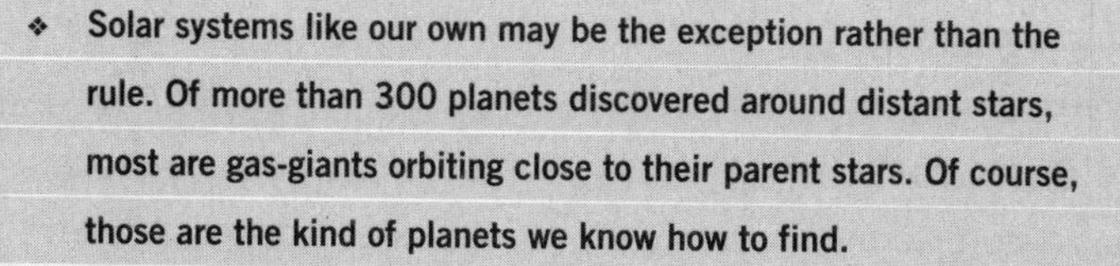

- Solar systems like our own may be the exception rather than the rule. Of more than 300 planets discovered around distant stars, most are gas-giants orbiting close to their parent stars. Of course, those are the kind of planets we know how to find.
- The line between star and planet is a little fuzzy. In 2006, astronomers discovered protoplanetary disks around half a dozen planetlike objects, ranging in mass from 5 to 15 Jupiters. Researchers haven't yet agreed whether these represent wayward planets or "brown dwarfs," protostars that were too small to achieve significant fusion.

MERCURY

THE BASICS

Mercury is the smallest planet in the solar system, only 40 percent larger than Earth's moon, and the closest to the sun. It's also the fastest. Named for the winged messenger god of Roman mythology, it orbits the sun every 88 days. Viewed from Earth, it swings back and forth across the sky about twice a year. Like our moon, Mercury is scarred with numerous impact craters, as it has virtually no atmosphere in which meteorites can burn up.

One day on Mercury lasts two Earth years, time enough for its dayside to get broiled to 800 degrees Fahrenheit. Temperatures on Mercury's night side plummet as low as –279 degrees. Ice might even hide in its deepest shaded craters.

Based on its size and mass, researchers estimate Mercury has a core of mostly iron that makes up at least 60 percent of its mass and 75 percent of its diameter. Mercury is the only other rocky planet in the solar system besides Earth that has a magnetic field, which implies that its outer core is molten and generates electric currents as it flows. As the molten layer cools, the inner solid core should grow. Solid iron is denser than liquid iron, so the overall core would actually shrink.

ON THE FRONTIER

The surface of Mercury bears testament to its shrinkage in the form of elongated cliffs or scarps. Researchers believe these are places where the surface has cracked, with one piece of crust sliding underneath another and pushing it upward. Other surface features suggest past volcanic activity. Craters contain smooth patches of rock that seem to be hardened lava. Both conclusions were supported by data from NASA's MESSENGER (Mercury Surface, Space Environment, Geochemistry and Ranging) spacecraft, which flew within 125 miles of the planet in January of 2008, the first Mercury flyby in more than 30 years. MESSENGER is the first spacecraft designed to orbit Mercury. It is scheduled to enter orbit on March 18, 2011.

COCKTAIL PARTY TIDBITS

- **Mapping of Mercury's surface reveals scars from 15 major impacts, including the 960-mile-wide Caloris Basin, where a possible asteroid strike seems to have lifted and cracked the crust on the opposite side of the planet.**
- **Going to Mercury? Pack your shades. The sun would appear 2.5 times larger there than on our planet.**

VENUS

THE BASICS

Once known as the Morning and Evening Star, Venus, the second planet from the sun, is the brightest object in the night sky after the moon. But despite being named for the Roman goddess of love and beauty, Venus is ugly on the inside. Those yellowy clouds that make it shine so brightly are made of droplets of sulfuric acid. On the surface, temperatures reach 900 degrees Fahrenheit—hot enough to melt lead—and the pressure is equivalent to being 3,000 feet under Earth's seas. Probes that landed on Venus lasted only a few hours.

Similar to Earth in size, mass, and composition, Venus is the victim of the strongest greenhouse effect in the solar system. Its thick dry atmosphere, rich in carbon dioxide, retains more of the sun's heat than Mercury does, making Venus hotter despite being twice the distance from the sun. Without the greenhouse effect, researchers think Venus's climate would be similar to Earth's. In fact, Venus may have once had water oceans, but they would have boiled away in the runaway heat.

Venus is our closest neighbor, and we've sent more than 20 spacecraft there since 1962. Researchers once thought of it as the best place to look for life in the solar system. We've since learned it's probably among the worst.

ON THE FRONTIER

The European Space Agency's Venus Express probe has orbited the planet since 2006, revealing a huge atmospheric vortex near the planet's south pole. NASA plans to send the Venus In-Situ Explorer, which will return to the surface of Venus to drill into the crust and take core samples below the weather-beaten surface. Venus has gained increased attention in recent years because it could serve as an extreme case study for the global warming now affecting Earth.

That doesn't mean we'll end up dry like Venus. Researchers think the sun first baked off Venus's water over millions of years, and then, without life or geological reactions to soak up carbon dioxide, the greenhouse effect took over.

COCKTAIL PARTY TIDBITS

- Venus's retrograde rotation (it rotates in the opposite direction as it orbits) means that on its surface, the sun rises in the west and sets in the east.
- The temperature on Venus seems to vary only with altitude. The Magellan probe spotted a reflective substance on the planet's highest peaks, similar in appearance to snow. Researchers have speculated it might be condensed tellurium, an element, or lead sulfide.
- Radar scans of the Venusian surface show some large craters but few small ones. Small meteorites would burn up in its thick atmosphere.

EARTH

THE BASICS

Our home is a planet of superlatives: Earth is the largest of the rocky planets, the densest planet in the solar system, and the only one where water can exist in liquid form on the surface. It's also the only planet known to harbor life. Oceans cover 71 percent of Earth's surface, absorbing the sun's heat and helping keep temperatures stable. The planet's iron–nickel core generates a strong magnetic field. Our dense atmosphere protects life from deadly cosmic rays.

Earth's atmosphere is notable for its oxygen (21 percent), which comes from respiring plants and photosynthetic microbes. The rest is mainly nitrogen (78 percent) along with traces of argon (almost 1 percent), water vapor, and, notably, carbon dioxide, which helps warm the surface by trapping infrared light. The rocky surface is relatively young, renewing itself regularly through the shifting of tectonic plates in the crust. The planet's axis of rotation is tilted 23.4 degrees from vertical, which leads to slight variations in temperature over the globe during the year that cause the seasons.

The history of life on Earth depends on how quickly it settled into a habitable place after forming 4.5 billion years ago. The earliest evidence for life—albeit pretty sketchy—comes from rocks in Greenland dated at 3.8 billion years old.

They have slightly more carbon-12 than carbon-13, suggesting the work of microscopic life.

ON THE FRONTIER

Scientists used to think that Earth was inhospitable to life for the first 700 million years, a period called the Hadean, Greek for Hell. Then, after a cometary bombardment delivered large quantities of ice, which melted and became our oceans, life must have evolved rapidly, which seemed a little suspicious. But in recent years, studies of small zircon crystals from Australia have revised that view. Studies of these 4.2 billion-year-old zircons—some of the oldest known rocks—suggest that Earth had oceans as well as continental plates as of 4.2 billion years ago, giving life more time to evolve.

COCKTAIL PARTY TIDBITS

- **Up to 90 percent of Earth's heat comes from radioactive decay of elements such as potassium-40, uranium-235, uranium-238, and thorium-232.**
- **The sky is blue because blue light from the sun scatters more than light of other colors as it hits gas molecules in the atmosphere. Viewed from space, the sun appears less blue and more orange.**
- **The continents shift by a few centimeters a year, or about the same rate that fingernails grow.**

THE MOON

THE BASICS

The most prominent body in the sky after the sun, the moon made such an impression on humanity that as soon as we got around to counting, we began to measure the passage of time based on its regular cycles (29.5 days for it to return to same spot in the sky). To date, it's the only place in the solar system besides Earth we've set foot on, and we'll probably return there before we go anywhere else.

Two main surface types give the moon its distinctive look: ancient, light-colored highlands (known as terrae) contrast with smooth dark plains (called maria), formed when massive craters flooded with lava. Comet impacts over time likely introduced some water to the lunar surface. Most of these water molecules probably get split by sunlight into oxygen and hydrogen, then ejected into space, but it's possible that ice lingers in some of the deepest craters near the poles, which never see sunlight.

Researchers believe the moon formed some 4.5 billion years ago, when an asteroid or small protoplanet struck Earth and broke off a piece of our still-forming planet. With practically no atmosphere or magnetic field to protect it from the ravages of space, the moon became an easy target for meteorites. Of the craters dotting the moon, more than a million are half a mile across or larger.

ON THE FRONTIER

Although we last visited the moon in 1972, NASA announced plans in 2006 to investigate building a permanent base near the poles, where scientists and astronauts could investigate natural resources, conduct extended experiments, and prepare for future missions to Mars. The United States will launch the Lunar Reconnaissance Orbiter (LRO) no earlier than summer 2009, to pave the way for a lunar base by as early as 2020.

LRO will carry a second craft, the Lunar Crater Observation and Sensing Satellite (LCROSS), designed to crash into the moon's south pole to uncover any hidden water ice.

COCKTAIL PARTY TIDBITS

- **The moon drifts 1.6 inches farther from Earth every year, an effect of the ocean tides. Earth's rotation slows by a small amount for the same reason. Scientists think an Earth day used to be some five or six hours long. So blame the moon for the current length of the workday.**
- **The moon has one of the largest impact craters in the solar system. The South Pole–Aitken basin, situated on the far side of the moon in its southern hemisphere, measures 2,240 kilometers across and 13 kilometers deep.**

MARS

THE BASICS

Mars, the fourth planet from the sun, is nicknamed the Red Planet for its ruddy hue, visible even to the naked eye. This coloration comes from oxidized iron minerals—literally, rust—in its surface rocks and soil. Powerful Martian winds only deepen the redness when they blow soil into the thin dry atmosphere.

By planetary standards, the Martian climate is relatively mild. Surface temperatures average a chilly –86 degrees Fahrenheit, with lows of –200 degrees in the winter, but summer days can get as warm as 80 degrees.

If Mars had national parks, the price of admission would be high. The Martian landscape is host to both the tallest mountain and the deepest canyon in the solar system. Olympus Mons, an inactive volcano, stands 16.7 miles high and spreads 342 miles across. The Valles Marineris trench extends across the Martian surface for nearly 2,500 miles and cuts almost 4 miles deep.

Ice caps at both poles hint that Mars had a wet past. Dry riverbeds and ancient floodplains support the idea of a prolonged period of widespread liquid water some three to four billion years ago. Today the planet is too cold and dry to support liquid water on the surface, but researchers still hope they might find it buried, perhaps with evidence of past life.

ON THE FRONTIER

Mars has more robotic visitors on or around it than any other planet in the solar system save Earth. The Mars Explorer rovers Spirit and Opportunity, still puttering around the dusty surface since landing in 2004, have supplied compelling evidence that Mars had large bodies of standing water in the past; Mars Reconnaissance Orbiter snapped images in 2007 of suspicious snaking avalanches flowing down gulley walls (maybe water, maybe fine dust); and Mars Phoenix Lander set down near the north pole in 2008, where it confirmed the presence of ice beneath the soil. Next up: Mars Science Laboratory. Scheduled to launch in 2011, this souped-up rover is equipped with a laser to scan the chemical makeup of rocks from a dozen yards away.

COCKTAIL PARTY TIDBITS

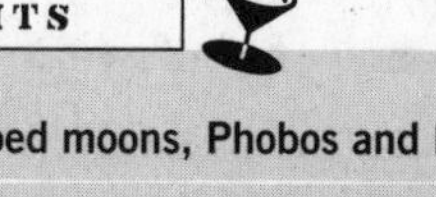

- **Mars has two irregularly shaped moons, Phobos and Deimos, which are likely wayward asteroids pulled in from the nearby asteroid belt. Within 30 to 80 million years, Phobos will either impact the Martian surface or, more likely, break up to form a planetary ring.**
- **A giant asteroid seems to have slammed into Mars more than four billion years ago, leaving a 20-mile-deep elliptical depression (the "hemispheric dichotomy") covering 42 percent of the planet.**

ASTEROIDS

THE BASICS

Asteroids are the crumbs of the solar system: dry dusty pieces of rock and iron floating in space. Most known asteroids orbit the sun in a belt between Mars and Jupiter, where they constantly collide. Astronomers have discovered nearly 300,000 of them, but they predict there are a billion or more out there. They range in size from Ceres, which is round and measures 580 miles across, to as small as a few city blocks. Other asteroids probably started off round, but they have been ground down to smaller sizes. All together, their mass is a fraction of the moon's.

Earth shows multiple scars from asteroid impacts, such as the Chicxculub crater, a 110-mile-wide formation on the Yucatan coast that researchers blame for the death of the dinosaurs. Researchers believe it came from an asteroid six miles wide that struck Earth 65 million years ago, kicking up a cloud of dust that darkened skies and cooled the globe.

Thousands of asteroids cross or come close to Earth's orbit. Skywatchers have identified nearly 1,000 of these near-Earth objects (call them NEOs to fit in with space buffs) as potentially hazardous asteroids ("PHAs"), defined as larger than 500 feet across and coming within 4.7 million miles of Earth's orbit. Researchers hope that by monitoring them, they can predict the threat of an impact.

ON THE FRONTIER

The NASA spacecraft Dawn is on an eight-year mission to rendezvous with the asteroid Vesta in 2011 and the dwarf planet Ceres in 2015. These are two of the biggest asteroids in the belt between Mars and Jupiter. Dawn is designed to study their shape, geological history, and composition, including the presence of water-bearing minerals. If asteroids are indeed the seeds of planet formation, as researchers are pretty sure they are, the study ought to give a better picture of what those early planet crumbs were like.

COCKTAIL PARTY TIDBITS

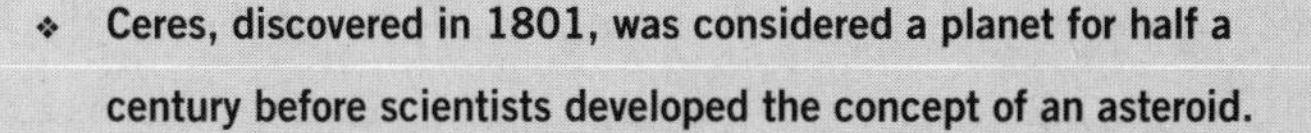

- **Ceres, discovered in 1801, was considered a planet for half a century before scientists developed the concept of an asteroid.**
- **NASA claims it found at least 168 impossible things in screenings of the asteroid doomsday film *Armageddon*, which it uses for management training. Among them: the asteroid wouldn't have Earthlike gravity, and you couldn't land a shuttle on it anyway.**

METEORS

THE BASICS

Meteoroid is a catchall term for smaller space rocks, including fragments of asteroids or comets, or even the rare rock blasted loose from Mars or the moon. Once broken off, each meteoroid begins its own highly irregular orbit around the sun. When one of these rocks streaks through Earth's atmosphere with a flash of light, it's called a meteor. If it survives the journey and falls to Earth, it's called a meteorite.

Comet remnants often orbit together, creating brilliant meteor showers when they cross into our atmosphere. The Perseid meteor shower occurs every year between August 9 and 13 when the Earth passes through the orbit of the comet Swift–Tuttle. Halley's comet is the source of the Orionid shower in October.

Meteors are usually seen 40 to 75 miles up in the air, and they can collide with the Earth at speeds of up to 44 miles per second. When such a rock enters the upper atmosphere, it ionizes molecules in the air leaving a bright streak known as an ionization trail. Although these bright flashes can last up to 45 minutes, most meteors burning up in the atmosphere are sand-grain sized and don't create any fireworks at all.

Despite the lack of display, we're under constant bombardment: it's estimated that these tiny meteorites enter the atmosphere every few seconds or so.

ON THE FRONTIER

Meteoroids can make you ill. In September 2007, a meteor lit up the sky and left a crater 60 feet wide and 15 feet deep near Lake Titicaca in Peru. Within days, townspeople started getting sick with symptoms of severe dizziness, skin lesions, nausea, and vomiting. Deadly space microbe? Nope. It turned out that the rocky meteor vaporized part of an underground water supply laced with arsenic. Carried by the wind, the arsenic-laced steam spread to bystanders and local townspeople and made them ill.

COCKTAIL PARTY TIDBITS

- **Small meteorites can be a real threat to spacecraft. The Hubble Space Telescope has 572 craters and chips from meteorite impacts. There's a guy at NASA whose job is to simulate these impacts by shooting fake meteorites out of a big gun.**

JUPITER

THE BASICS

More than twice as massive as all the other planets combined, Jupiter is about as big across as planets can get; adding more material would only compress it, making the planet denser while barely increasing its diameter. It is composed of the primordial elements of the solar nebula that created the sun: 90 percent hydrogen and 10 percent helium. If Jupiter's mass were 80 times greater, hydrogen nuclei in its core would undergo nuclear fusion, causing the planet to expand exponentially in size and energy to become a star.

If you voyaged deep into Jupiter's guts, where temperatures rise as high as 20,000 degrees Fahrenheit and pressures reach four million times that of Earth, you'd find an elemental rarity: liquid metallic hydrogen. This happens only when hydrogen atoms are ripped apart so that electrons flow freely. The resulting electrical currents generate a huge magnetic field that extends past Saturn.

Only Jupiter's uppermost layers of hydrogen and helium are visible on its roiling surface. Its extreme rotation (a Jovian day lasts 10 hours) whips up powerful turbulence that breaks its atmosphere into parallel bands of storms. Most famous of these storms is the Great Red Spot, which is big enough to swallow four Earths and has lasted perhaps 300 years.

ON THE FRONTIER

Jupiter is a mini-planetary system unto its own. It has 63 known moons, most named for the myriad lovers of Zeus, or Jupiter as the Romans called him. They vary wildly in character: violent Io has active volcanoes that spew sulfur plumes hundreds of miles into space; Callisto is the most pockmarked object in the solar system; and Ganymede is the largest moon in the solar system, bigger than Mercury. Most captivating may be Europa. Researchers believe it harbors a vast water ocean beneath its icy smooth surface, which is riddled with brown fractures. Geothermal heat from inside the moon could provide an excellent breeding ground for life. NASA is already testing a $2.3 million submersible to potentially have a look-see in 2028.

COCKTAIL PARTY TIDBITS

- **Jupiter has yet to finish cooling down from its formation. It puts out about 70 percent more heat than it gets from the sun, from trapped infrared radiation.**
- **Even if the Great Red Spot fades, future generations might have a new superstorm to track. In 1999, NASA discovered a "large white spot" about half the size of the Great Red Spot that formed from a conglomeration of smaller storms. By 2006, the growing storm had turned red.**

SATURN

THE BASICS

Few extraterrestrial sights spark as much awe as Saturn's bright swath of rings. Throw in a nearly flawless complexion—pale yellow and creamy—and Saturn is the most photogenic planet in the solar system by far. But this planet has more going for it than poster-quality good looks. Composed of the same material as Jupiter, Saturn spins so fast that it bulges at the equators and flattens at the poles, forming an oblong shape. It is the second-largest planet in the solar system but also the least dense. Saturn would float in water, if you could find an ocean large enough to dunk it.

Like Jupiter, Saturn has numerous satellites (60 confirmed) and a liquid metallic hydrogen interior surrounding a small rocky core. It generates similar cloud bands and storm spots, but they're much more difficult to distinguish in its custardlike surface, the product of smoggy ammonia.

Saturn's famous rings are hundreds of thousands of miles across but only a few meters thick. Composed mostly of microscopic particles of water ice (although some are as big as cars), they are held in place by the gravity of the planet and a few "shepherd" moons. The rings may have formed from a moon that disintegrated, perhaps as recently as the time of the dinosaurs.

ON THE FRONTIER

The most fascinating of Saturn's moons is Titan, the second-largest in the solar system after Jupiter's Ganymede. Larger than Mercury and denser than Pluto, it's the only moon in our solar system with a significant atmosphere and the only place in the solar system other than Earth with liquid on its surface. When NASA's Cassini-Huygens mission scanned beneath Titan's hazy orange skies in 2004, it uncovered not a methane ocean, as many researchers had speculated, but lakes of liquid hydrocarbon at the moon's poles. More recent findings indicate that Titan may have a liquid water ocean 30 to 90 miles beneath its surface.

COCKTAIL PARTY TIDBITS

- The rings of Saturn actually have an atmosphere of sorts, made of molecular oxygen that results when ultraviolet light from the sun reacts with water ice in the rings.
- Like Titan, Saturn's moon Enceledus is also thought to conceal water beneath its icy surface. Recent observations suggest that frosty volcanoes on its surface regularly eject ice crystals into space, where they may rejuvenate and replenish Saturn's rings.
- Saturn generates as much heat as it gets from the sun, by a strange mechanism: friction from droplets of liquid helium falling through the planet's metallic layer.

URANUS

THE BASICS

Uranus is the third-largest planet in the solar system. Unlike Jupiter and Saturn, hydrogen makes up only 15 percent of Uranus's mass, with helium present in trace amounts. Both elements are concentrated in the atmosphere, which glows a pale blue because high methane clouds scatter light in much the same way as Earth's atmosphere does. At the center of the planet lies a rocky icy core not much larger than Earth, surrounded by dense layers of water, methane, and ammonia.

Uranus radiates very little heat and receives very little warmth from the sun (one 400th the intensity of sunlight on Earth), making it the coldest planet in the solar system. Its atmospheric temperature drops as low as –370 degrees Fahrenheit. Some scientists think an early collision with an Earth-mass protoplanet might have stripped Uranus of most of its warm core.

This same collision could be responsible for the planet's oddest feature: its axis of rotation is tilted on its side, with one pole pointed toward the sun. Whereas all the other planets spin like a top, Uranus spins like a rolling ball. Each pole gets 42 years of continuous sunlight, followed by 42 years of darkness.

ON THE FRONTIER

Uranus was the second planet to show astronomers its rings, demonstrating they are a common planetary feature. The planet's 13 known rings are dark and made of particles of dust and rock that may have come from a shattered moon. Recent research indicates that the rings have changed dramatically even since the Voyager 2 space probe observed them in 1986. Telescopes aimed at the planet in 2007 showed that one of Uranus's major rings has migrated thousands of miles out from its original position.

COCKTAIL PARTY TIDBITS

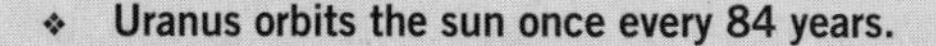

- Uranus orbits the sun once every 84 years.
- Uranus is sometimes just barely visible with the naked eye on exceptionally clear, dark nights. When astronomer William Herschel first identified it in 1781, he thought it was a comet.
- Uranus has 27 known moons, 26 of which are named for Shakespearean characters (Miranda, Titania, Oberon, etc.). The rest of the moons in the solar system are named for Greek gods.

NEPTUNE

THE BASICS

Like the Roman sea-god for whom it was named, Neptune, the eighth and final planet in the solar system, is a tempestuous place, home to the fastest winds in the solar system, up to 1,300 miles per hour. As massive as 17 Earths, Neptune is slightly larger than Uranus and denser, making it the third most massive planet in the solar system.

During Voyager 2's 1989 flyby, cameras observed a storm thousands of miles wide, called the "Great Dark Spot," near the equator as well as a smaller dark spot to the south. When the Hubble Space Telescope checked again in 1994, the Great Dark Spot was gone, replaced by a similar storm in the northern hemisphere.

Like Uranus, beneath Neptune's atmosphere lies a mantle of water, methane, and ammonia surrounding a small rocky core. Called ice by scientists, the mantle is actually a dense, electrically conductive liquid. Researchers believe Neptune has more varied weather than Uranus because it generates more internal energy, by unknown mechanisms.

Neptune orbits the sun once every 165 years. In 2011, it will finish its first complete orbit around the sun since it was discovered in 1846. It has 13 moons and a faint ring system, which might appear red up close due to carbon or silicate dust coating the icy ring particles.

ON THE FRONTIER

Orbiting the sun at a distance of 2.8 billion miles, or 30 times Earth's distance from the sun, Neptune is impossible to see with the naked eye. Astronomers deduced its presence in 1846 from irregularities in the orbit of Uranus. But even at such a great distance from the sun, small variations in solar heating may give rise to seasons. Neptune has grown steadily brighter since 1980, and over the past decade, its cloud bands have grown wider and brighter as well. If Neptune is truly undergoing seasonal change, it should continue to brighten over the course of our lifetimes.

COCKTAIL PARTY TIDBITS

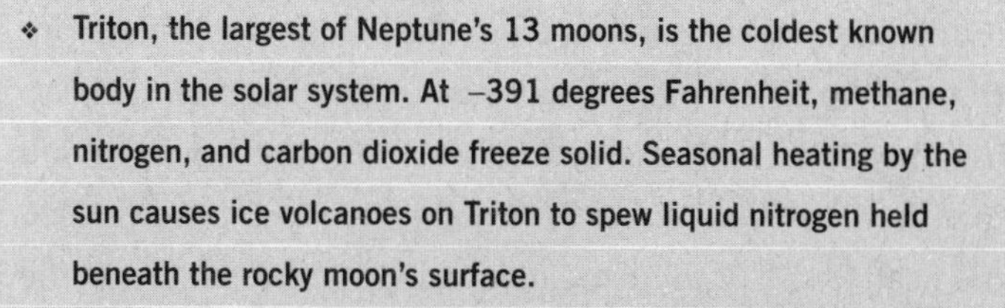

- **Triton, the largest of Neptune's 13 moons, is the coldest known body in the solar system. At –391 degrees Fahrenheit, methane, nitrogen, and carbon dioxide freeze solid. Seasonal heating by the sun causes ice volcanoes on Triton to spew liquid nitrogen held beneath the rocky moon's surface.**
- **Methane escaping from Neptune's poles floats high enough in the atmosphere to form wispy white clouds that don't look too different from cirrus clouds on Earth. But on blustery Neptune, those clouds move faster than the speed of sound.**

PLUTO AND THE KUIPER BELT

THE BASICS

Pluto, discovered in 1930, was long considered the ninth planet. But it was always an outcast. Made of frozen gases and rock, Pluto is 1,400 miles wide, or just two-thirds the size of the moon. And its orbital path around the sun is so tilted that it crosses Neptune's orbit for 20 years out of every 249-year revolution. (From 1979 to 1999, it was closer to the sun than Neptune.)

Researchers now believe Pluto is one among tens of thousands of frozen cometlike rocks stably orbiting the sun past Neptune, in a region called the Kuiper (rhymes with "viper") Belt. Since 1992, researchers have discovered some 70,000 Kuiper Belt objects larger than 100 kilometers across, including the dwarf planets Haumea and Makemake.

The Kuiper belt extends 55 times farther from the sun than Earth. Beyond it lies yet another group of rocks, called the scattered belt, which extends about twice as far as the Kuiper Belt. Scattered belt objects have more erratic orbits and may be the source of some comets.

In 2003, astronomers discovered Eris, a scattered belt object 5 percent bigger than Pluto, which was briefly called the tenth planet. Prompted by the discovery, the International Astronomical Union voted to demote Pluto and Eris to a new category, dwarf planets, now called "plutoids."

ON THE FRONTIER

It's hard to glean details about Pluto. It's so small and distant that even the Hubble Space Telescope can detect only contrasting light and dark areas on its surface, which researchers speculate might come from uneven concentrations of frozen gases such as carbon and nitrogen. To find out more, NASA launched the New Horizons mission in early 2006. Traveling at 50,000 miles per hour, the space probe is expected to reach Pluto in 2015 to snap images and gather detailed data about Pluto's composition, the geological forces that shape its surface, and its atmosphere.

COCKTAIL PARTY TIDBITS

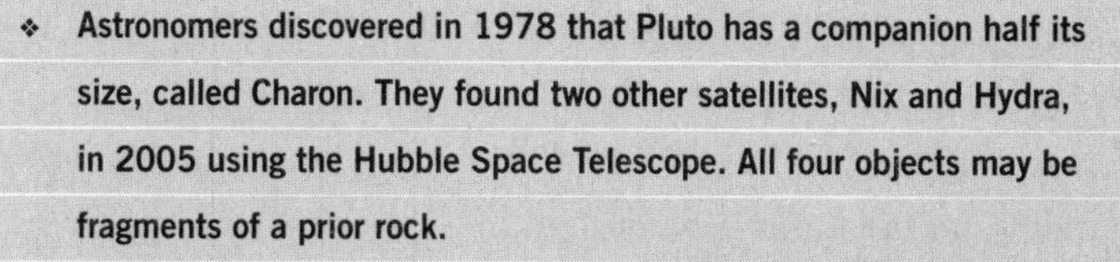

- **Astronomers discovered in 1978 that Pluto has a companion half its size, called Charon. They found two other satellites, Nix and Hydra, in 2005 using the Hubble Space Telescope. All four objects may be fragments of a prior rock.**
- **Eris was originally named Xena by its discoverers, after the main character of the television show *Xena: Warrior Princess*, played by actress Lucy Lawless. They renamed it Eris (after the Greek goddess of strife), but its moon remains connected to the show: *Dysnomia* means "lawless."**

COMETS

THE BASICS

Comets are the drifters of the solar system. Leftovers of the raw material that formed the planets, they are loose, irregularly shaped clumps of ice, dust, and small rocks, measuring up to tens of miles across.

Most of the time, comets resemble asteroids. (In fact, some asteroids may be dead comets.) But when they pass within a few hundred million miles of the sun, solar radiation vaporizes ice on the comet's surface into a cloud up to 60,000 miles across and blue tails that may extend up to 100 million miles.

Some comets take only a few years to orbit the sun. These objects may come from the Kuiper Belt or the scattered cloud beyond it, groups of icy rocks past Neptune. Collisions between comets or the gravity of the outer planets may nudge them into the inner solar system.

Comets often travel in highly elliptical orbits that bring them close to the sun for a portion of their orbit, or they may only pass through once on their way out of the solar system. Researchers believe these infrequent visitors come from an even more distant region known as the Oort Cloud, 100 times farther away than the Kuiper Belt, where comets are subject to the gravitational pull of neighboring stars.

ON THE FRONTIER

Comets have been seen in the sky and interpreted as omens since ancient times, and we still know very little about them, such as how much ice they contain. They are among the darkest objects in space. NASA's Deep Space 1 probe flew past Comet Borrelly in 2001, confirming that its surface absorbed more light than asphalt does. In 2005, NASA's ongoing Deep Impact mission fired a washing machine–size impactor into comet Tempel 1. The explosion, equivalent to five tons of TNT, ejected more than 1,000 tons of comet material into space. Researchers concluded the comet was about 75 percent empty space, with the consistency of lemon meringue. Mmm . . . comet meringue.

COCKTAIL PARTY TIDBITS

- Comets often collide with planets and other orbiting bodies. Comet Shoemaker-Levy broke into hundreds of pieces before smashing into Jupiter in 1994. Scientists think a fragment of Comet Encke caused the Tunguska event in 1908, which flattened trees and obliterated an 830-square-mile patch of Siberia.
- Comets lose mass every time around, about six feet of material per year. Comet Borrelly, which orbits the sun every seven years, is two miles across and if it continues on its path will dwindle to nothing in 6,000 years.

CHAPTER

SIX

THE SECRET LIVES OF THE STARS

STARS

THE BASICS

Stars look like points in the sky, but astronomers can figure out the age, mass, and composition of stars by studying their motions through space and the light coming from them. Spreading starlight through a prism reveals which elements are present in the star, soaking up light of different wavelengths. That was the original way of classifying stars.

Today astronomers know that most of a star's properties boil down to mass. The bigger a star, the hotter and faster it burns. The coolest stars glow red, medium temperature stars are white and yellow, and the hottest and most rare stars shine blue or bluish white. Bigger stars give off more light, so blue stars are brightest, and red stars are dimmest.

As long as the stellar core is fusing hydrogen into helium, the star is prevented from contracting under its own weight. Most stars will spend 90 percent of their lifetimes—their adulthood, basically—burning hydrogen. Astronomers say these stars are on the "main sequence."

The biggest stars, blue supergiants, live fast and die hard, burning through their nuclear fuel in about a million years. The smallest stars, called red dwarfs, burn low and slow for tens to hundreds of billions of years. Smaller stars are born more frequently than larger ones. Factoring in their longer lifespans, they vastly outnumber bigger stars.

ON THE FRONTIER

Knowing how stars break down into different categories, astronomers can deduce the distance to the stars they find in their telescopes. If they can figure out a star's mass, they know how much light it must produce, so its apparent brightness can tell them how far away the star is. The intrinsic or absolute magnitude of a star is defined as how bright it would appear on Earth from a distance of 10 parsecs, or 32.6 light-years.

Astronomers calculate distances by looking for stars we think always have the same intrinsic brightness, called standard candles. Like a 60-watt light bulb, a standard candle will fade in a predictable way the farther away it gets. Examples are Cepheid variables, a type of star that expands and contracts in a regular way, and exploding stars called type 1a supernovae, which seem to always release the same amount of energy.

COCKTAIL PARTY TIDBITS

- **Stars appear to twinkle because of turbulence in Earth's atmosphere. Astronomers have to correct for these distortions in any ground-based telescope lens larger than about 10 centimeters across.**
- **Most stars are somewhere between one billion and 10 billion years old. The oldest star yet discovered, HE 1523-0901, is an estimated 13.2 billion years old, making it almost as old as the universe itself.**

NEBULAS

THE BASICS

In sci-fi shows, spaceships are always flying into a nebula. But they never tell you what nebulas are. We will. You see, only one tenth of the atoms in the galaxy are locked up in stars and their planets. The other 90 percent makes up a thin haze of gas and dust called the interstellar medium. The visible parts of this medium are vast clouds called nebulas that can stretch for thousands of light-years.

Stars are born from the interstellar medium, specifically from clouds of cold hydrogen molecules. When it gets the right trigger, such as the tug of a passing star or the blast wave of a supernova, the cloud fragments into smaller blobs that condense to form stars. As long as the piece of cloud has at least 8 percent of the mass of the sun, it will condense over millions of years into a dense ball capable of fusing hydrogen into helium. Balls of gas that don't meet the 8 percent cutoff are doomed to become brown dwarfs, "failed stars" that may burn weakly for tens of millions of years.

As stars form in the hydrogen clouds, they give off ultraviolet light, which ionizes the surrounding gas and causes it to glow visibly, providing some of the most beautiful views in astronomy. These bright nebulas are called HII regions or emission nebulas.

ON THE FRONTIER

As stars age and die, they give back to the interstellar medium, either by shedding their outer envelopes into space, in the case of red giants, or blasting their matter far and wide when they go supernova. These outbursts enrich the elemental content of the next generation of stars. When stars form today they are composed of about 70 percent hydrogen and 28 percent helium, as measured by mass, with a small fraction of heavier elements. Over time, the fraction of helium in newborn stars will grow. Researchers estimate that eventually stars will start off their lives consisting of as much as 60 percent helium.

COCKTAIL PARTY TIDBITS

- **Emission nebulas are the visible tips of the surrounding hydrogen clouds, which absorb passing light, making them too dark to see unless viewed against the backdrop of another nebula.**
- **Supernovas put out a lot of dust—sooty molecules made of carbon, silicon, and oxygen—that hangs out in space blocking astronomers' views. According to a 2008 estimate, if all the dust were swept aside, the universe would be twice as bright as it is now.**
- **Streaks and other structures visible in nebulas may be the product of magnetic fields twisting through space, carried by stellar wind from hot stars.**

RED GIANTS

THE BASICS

When a star runs out of hydrogen fuel in its core, the path it takes next depends on its mass. For stars like our sun, their fate is to expand into a mighty red giant, hundreds of times its original size but much cooler. This transformation happens because the star begins burning hydrogen located in a shell around the core.

The newly burning hydrogen expands the star's outer envelope, which cools to a temperature of around 9,000 degrees Fahrenheit and begins to leak away, forming a planetary nebula. A red giant may lose up to a third of its mass this way. Some red giants will contract again when helium in the core begins fusing into new elements and then re-expand when the new fuel source runs out. Depending on the mass of a star, it may go through multiple red giant phases.

Stars more massive than about 10 suns expand into red supergiants, up to 1,500 times the size of the sun, making them the largest stars in the universe by volume, although not the most massive. Examples of red supergiants include the stars Betelgeuse and Antares. The smallest stars never make it to the red giant phase. When they use up all their hydrogen, they gradually cool and contract into red dwarfs.

ON THE FRONTIER

The red giant phase spells doom for any planets orbiting close enough to be engulfed by the bloated star. Scientists agree that when the sun goes red giant and reaches its maximum size 7.6 billion years from now, it will extend past the current orbit of Earth. They are less sure where Earth will be. The sun will drop about a third of its mass by that time, which may cause Earth's orbit to expand. But the shed solar material may exert a drag on Earth that would make it drift closer. Either way, life wouldn't survive. Radiation from the expanded sun would fry Earth to a crisp.

COCKTAIL PARTY TIDBITS

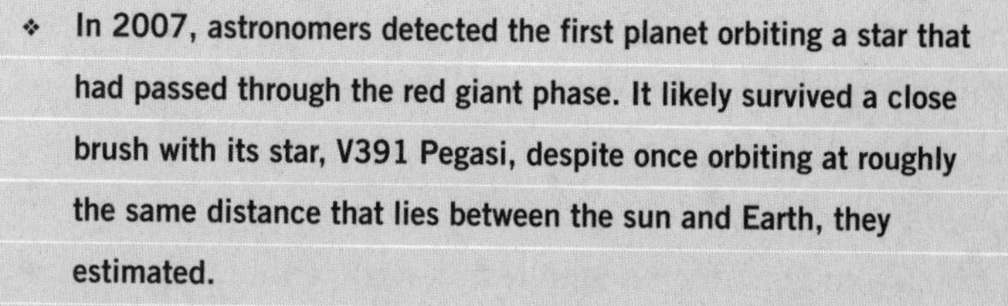

- In 2007, astronomers detected the first planet orbiting a star that had passed through the red giant phase. It likely survived a close brush with its star, V391 Pegasi, despite once orbiting at roughly the same distance that lies between the sun and Earth, they estimated.
- The nearest red supergiant is Betelgeuse (sometimes pronounced "beetle juice," like the movie), 600 light-years away. It is the star that appears largest in the sky after the sun.

WHITE DWARFS

THE BASICS

As a star goes through the red giant phase, it expels most of its mass in a cloud called a planetary nebula. The core of the star remains, but its mass has been squeezed into a much smaller space, transforming into a hot dense star called a white dwarf. White dwarfs typically have half the mass of the sun but are only slightly bigger than Earth.

When the core stops burning nuclear fuel, gravity begins to squeeze together its atoms until something called degeneracy pressure stops the process. Above and beyond electrons' mutual electric repulsion, they simply refuse to share the same space, which eventually halts the collapse of a white dwarf. All that squeezing heats the dwarf to a temperature of hundreds of thousands of degrees.

A funny consequence of degeneracy pressure is that the more massive a white dwarf is, the smaller it has to be for its electrons to generate enough resistance to counteract gravity. But if a white dwarf grows too massive, its atoms will undergo yet another transformation. The mass limit is 1.4 solar masses, referred as the Chandrasekhar limit, after Indian physicist Subrahmanyan Chandrasekhar.

White dwarfs make up about 6 percent of nearby stars, but they are hard to detect because they don't emit very much light. Researchers believe some 97 percent of the stars in the Milky Way will end up as white dwarfs.

ON THE FRONTIER

White dwarfs hang on to their heat pretty tightly. The electrons crushed together inside them should conduct heat very well, but researchers believe the surface of a white dwarf consists of a crust of carbon and oxygen some 30 miles thick. This surface layer would keep energy from escaping as radiation. Over many billions of years, a white dwarf should gradually redden and fade into a black dwarf. But even 13.7 billion years after the Big Bang, researchers are pretty sure even the oldest white dwarfs still have a temperature of thousands of degrees.

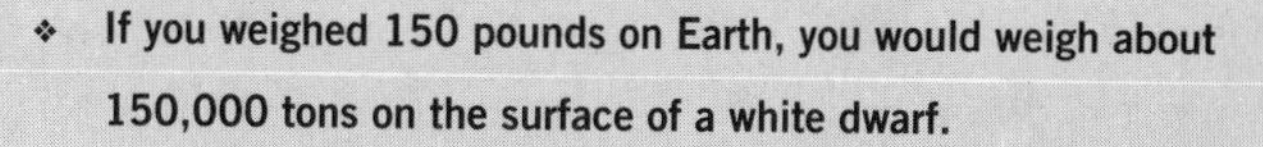

COCKTAIL PARTY TIDBITS

- **If you weighed 150 pounds on Earth, you would weigh about 150,000 tons on the surface of a white dwarf.**
- **Most white dwarfs should be made of carbon and oxygen, because the types of main-sequence stars that give rise to them would not typically be large enough to create any heavier elements.**
- **White dwarfs have atmospheres of nearly pure hydrogen or helium. (Heavier elements would mostly sink below the surface.) But the star's intense gravity would restrict these atmospheres to the height of an earthly skyscraper.**

SUPERNOVAS

THE BASICS

Finally, here's what you've been waiting for: explosions. When a big enough star runs out of nuclear fuel, its core collapses violently, producing a huge explosion called a supernova. Over a span of a few minutes, a supernova can briefly outshine whole galaxies, shedding as much energy as the sun during its entire lifetime. It then fades over a period of weeks or months.

When the core collapses, it sends out a shock wave that rips apart the outer envelope of the star and sends it flying at several tenths of the speed of light. By the time a star has reached this advanced age, it's often got a cloud of gas around it that leaked from the star. The shock wave plows into this material, generating lots of visible light, ultraviolet, and x-rays. The gas and dust keep expanding in what's known as a supernova remnant.

There are different types of supernova, depending on what kind of star is exploding. Type II supernovas (or supernovae, if you're feeling pretentious) sound the death of stars heftier than about eight suns. Type I supernovas happen when a white dwarf sucks up matter from a nearby star, triggering a runaway nuclear reaction.

Supernova remnants enrich the interstellar medium, the gas and dust between stars, with heavy elements that serve as the raw materials for planets. The expanding shock wave can also trigger a new star formation when it passes through a hydrogen cloud.

ON THE FRONTIER

Astronomers usually detect supernovas by their visible afterglows. They had long believed this glow was preceded by a flash or "outburst" of x-rays as the exploding star slammed into surrounding gas. In 2008, researchers finally caught a supernova red-handed and confirmed this model when they observed an explosion of x-rays 88 million light-years away. A few months later, other scientists witnessed another supernova prepping for its outburst, a bright glow of ultraviolet indicating that the star surged in temperature before it exploded.

COCKTAIL PARTY TIDBITS

- **Researchers estimate that about three supernovas should pop off per century in the Milky Way, but many of them may be hidden from view.**
- **In 2006, astronomers observed the brightest supernova ever. Called 2006gy, it burned 100 times brighter than a typical supernova for an amazing three months, and was still going strong after eight months. The exploding star may have weighed as much as 100 suns.**

NEUTRON STARS AND PULSARS

THE BASICS

After a star goes supernova, its core will collapse into something more exotic than a white dwarf. A mass of up to two stars is squeezed into a ball 12 miles across, a state so dense that protons and electrons fuse into neutrons. The result is a neutron star. Far denser than a white dwarf, one teaspoonful of neutron star material would weigh a billion tons.

Neutron stars are most commonly observed as pulsars, steady sources of radio waves or other radiation that pulse at regular intervals, like a lighthouse. Pulsars exist because neutron stars rotate rapidly and possess powerful magnetic fields that accelerate charged particles in twin streams at near the speed of light. The streams project from the magnetic poles of the neutron star. But as with Earth, the magnetic poles can be offset from the axis of rotation, causing the beams to twirl like jets of water from lawn sprinklers.

Pulsars can rotate hundreds of times a second. In 1982, scientists discovered a pulsar with a rotation period of only 1.6 milliseconds. But as pulsars age, they gradually slow to one rotation every few seconds.

Some pulsars emit x-rays. The famous Crab Nebula, a remnant born from a supernova observed in 1054 A.D., contains an x-ray pulsar at its core.

ON THE FRONTIER

Astronomers have found about 2,000 neutron stars, including 1,500 pulsars, usually in supernova remnants but sometimes by themselves, or sometimes in groups of two or more orbiting one another. The closest known neutron star, PSR J0108-1431, lies about 280 light-years away. About 5 percent of neutron stars are found in binary systems, orbiting stars, white dwarfs, other neutron stars, or even black holes.

Twin pulsar systems—one pulsar orbiting another—give researchers a unique tool for testing Einstein's general theory of relativity in cases where gravity is really powerful. They think these systems should be a major source of gravitational waves, ripples in space-time propagating at the speed of light.

COCKTAIL PARTY TIDBITS

- **Pulsars rock: For the cover of their 1979 debut album, *Unknown Pleasures*, English rock band Joy Division features an image of radio waves from pulsar CP 1919.**
- **The first planets discovered outside the solar system were found around pulsars. As of 2008, there were five confirmed pulsar planets, believed to be either the rocky cores of former gas giants or accumulated from trace solids left over from a supernova.**

GAMMA RAY BURSTS

THE BASICS

If supernovas aren't powerful enough for you, try gamma ray bursts. About once a day, at a random location in the sky, a distant point shines intensely with gamma rays, the most energetic type of electromagnetic radiation. These gamma ray bursts typically last a few seconds, but can be as quick as a millisecond or as long as two minutes. Following the burst for hours or days is an afterglow of radiation from x-rays to radio waves.

For a long time, scientists couldn't figure out what to make of these explosions, which are visible only from space because Earth's atmosphere absorbs gamma rays. Each one releases the combined energy of 1,000 stars like our sun, far more powerful than a typical supernova. Satellite observations indicate that bursts occurred in distant galaxies, anywhere from several billion to 13 billion years ago.

The leading model of gamma ray bursts holds they occur when the core of a massive, rapidly rotating star collapses, producing a gigantic supernova. To generate a gamma ray burst, an exploding star would have to weigh as much as 30 suns or more, and such a star would most likely collapse to form a black hole, an object so dense not even light can escape it. So researchers think gamma ray bursts are the birth cries of black holes.

ON THE FRONTIER

In 2003, researchers confirmed that the afterglow of a nearby burst had the same optical spectrum as a prior supernova observed in 1998. If it quacks like a supernova, then it must be a supernova, is how the thinking goes.

Still perplexing are bursts of a couple of seconds or less, which tend to come from older galaxies where powerful supernovas are less common. Researchers suspect the shorties happen when a pair of neutron stars collides and merges into one. NASA launched the Fermi Gamma Ray Space Telescope (formerly GLAST) in the summer of 2008 to canvas the sky for sources of gamma rays and to get a closer look at bursts.

COCKTAIL PARTY TIDBITS

- **The U.S. military satellite Vela 4 discovered the first gamma ray burst by accident in 1967, as it was monitoring for Soviet violations of the 1963 Nuclear Test Ban Treaty.**
- **Researchers estimate that only one out of every 100,000 supernovas is powerful enough to produce a gamma ray burst, which works out to about one burst per day. And that's what we see.**

BLACK HOLES

THE BASICS

When a star at least 10 to 15 times as massive as the sun explodes in a supernova, leaving behind a sufficiently massive stellar core or remnant, the result will be a black hole, an object so dense that not even light can escape it. A black hole forms when matter is squeezed into such a small volume that its gravitational attraction overwhelms any possible repulsion between atoms or subatomic particles. All the matter is crushed into a point of essentially zero volume and infinite density, called a singularity.

The gravity around a singularity is so strong that it pulls in everything passing within a certain distance, marked by the infamous event horizon. Whatever crosses that horizon doesn't come back, including light, hence the name black hole. If you were falling toward a black hole, the intense gravity around it would quickly tear you apart.

The distance to the horizon is equal to the Schwarzschild radius, which refers to how small you have to compress a star (or anything else) to turn it into a black hole. The Schwarzschild radius for the Earth is four-tenths of an inch. For the sun, it's about two miles, which means that if a mischievous alien race replaced the sun with an equally massive black hole, Earth would not be sucked in. It would just get really cold and dark.

ON THE FRONTIER

Black holes can actually put out a lot of light. An "active" black hole is one that's pulling matter around it in a spinning disk, which becomes very hot and gives off high-energy radiation. Using the orbiting Chandra X-ray Observatory, researchers can search for x-rays coming from these disks. They estimate there are probably 100 million stellar-mass black holes in the Milky Way, ranging in mass from 3 to about 100 suns. The closest candidate is 1,600 light-years away from Earth.

COCKTAIL PARTY TIDBITS

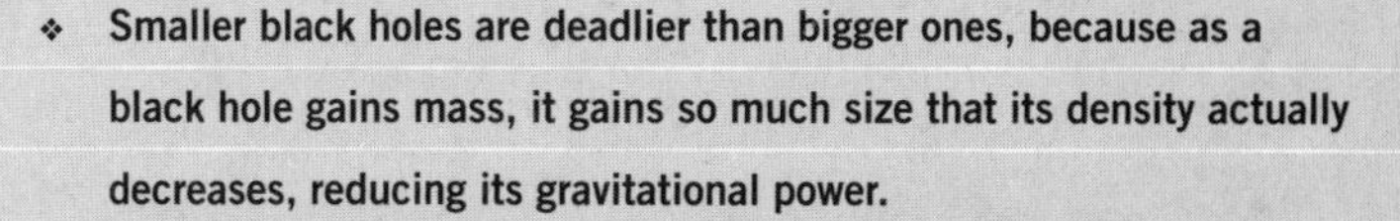

- **Smaller black holes are deadlier than bigger ones, because as a black hole gains mass, it gains so much size that its density actually decreases, reducing its gravitational power.**
- **Black holes can spin. They absorb the angular momentum of whatever matter they engulf.**
- **Nobody is sure what happens at a singularity because Einstein's theory goes haywire there. Researchers do believe that singularities are always surrounded by an event horizon; they're never "naked."**

CHAPTER

SEVEN

WEIRD MATTER AND ENERGY

ANTIMATTER

THE BASICS

Every subatomic particle has an evil twin, a sort of mirror-image counterpart that has the same mass but is otherwise its complete opposite. These are called antiparticles or antimatter. Antiparticles are produced in radioactive beta decay and whenever particles collide at high enough energy. When matter and antimatter meet, they annihilate each other in a burst of gamma rays and unstable particles. (*Star Trek* fans will recognize antimatter as the power source for the Starship *Enterprise*.)

The most obvious difference between a particle and its antiparticle is charge. The antiparticle of the electron is the positively charged positron. But neutrons and other electrically neutral particles also have antiparticles.

A universe made of antimatter would be nearly indistinguishable from our own. Positrons, antiprotons, and antineutrons would combine to make antihydrogen, antihelium, and other antielements. Antimatter scientists would discover the same forces of gravity, electromagnetism, and the rest.

Matter and antimatter destroy each other because when they come together, their opposite properties cancel each other out. For example, the negative charge on the electron and the positive charge on the positron add up to no charge. To the universe, the pair looks like pure matter, which, by

Einstein's equation $E=mc^2$, is equivalent to energy. By the same token, concentrated energy can generate particle–antiparticle pairs.

ON THE FRONTIER

CERN, the European particle physics lab near Geneva, produced the first antihydrogen atoms in 1995—nine of them—by bombarding xenon with antiprotons. Current methods yield about 100 antihydrogen atoms per second by bringing together positrons and antiprotons in a magnetic trap, but they quickly get annihilated. CERN recently estimated it would take two billion years to produce one gram of antihydrogen.

COCKTAIL PARTY TIDBITS

- **Because antimatter can be produced only in particle colliders, some have called it the most expensive substance on Earth. Antimatter could cost as much as $300 billion for one milligram.**
- **An antimatter bomb would be far more powerful than nuclear weapons. Matter–antimatter annihilation converts 100 percent of mass to energy, compared with 0.7 percent in a hydrogen bomb.**

COSMIC RAYS

THE BASICS

Cosmic rays are charged particles that are constantly raining down from explosions outside the solar system, packing a lot of energy. When they hit Earth's atmosphere, they explode in fireworkslike showers of high-energy particles. About 90 percent of all incoming cosmic rays are protons, 9 percent are helium nuclei, and 1 percent heavier nuclei, including rare elements and isotopes.

Cosmic rays get their energy from magnetic fields of galaxies and the sun, but these fields also scramble their flight paths, making their origins hard to trace. Some of them must come from the sun, because the number of incoming cosmic rays increases after solar flares. These rays move at 80 percent of the speed of light and would be potentially lethal to future astronauts voyaging to Mars or beyond.

Most cosmic rays probably come from supernovas, the explosions of dying stars, in our galaxy. Such explosions wouldn't literally shoot out cosmic rays. Instead they would send out an expanding cloud of plasma called a remnant, which lasts for thousands of years. Charged particles rattle around in the magnetic field of the remnant, and some of them pick up a big enough kick to break free.

ON THE FRONTIER

Short-lived particles called muons are produced when cosmic rays strike the atmosphere. About 10,000 muons fall on every square meter of Earth every minute. They can penetrate dozens of feet into the ground, unless dense elements block their path. Researchers at Los Alamos National Laboratory are developing a technology that uses muons to scan for illicit nuclear material. A sample is placed inside a muon-detection system that measures the speed and flight path of muons entering and exiting the detector, which tells them the possible composition of the sample. You still have to wait for enough muons to pass by, but at least there's no need for hazardous x-rays.

COCKTAIL PARTY TIDBITS

- **The most powerful cosmic rays pack the energy of a fast-pitch baseball, which dwarfs the particle energies we can create on Earth. Scientists suspect these ultra-high energy rays come from distant active galaxies, which are believed to harbor enormous black holes sucking in gobs of matter.**
- **The faint microwave glow in space should limit the energy of cosmic rays, because the rays would eventually start to scatter off of the microwaves.**

NEUTRINOS

THE BASICS

The neutrino is a ghostly elementary particle that has no electric charge, barely any mass, and moves at near the speed of light. A relative of the electron, it can pass through ordinary matter without leaving a trace. Tens of billions of neutrinos pass through every square inch of our bodies every second.

Neutrinos are produced in nuclear reactions in stars and in radioactive beta decay. When the universe was young and hot, these reactions took place in great numbers, and the neutrinos produced then are still buzzing around today. Those "relic" neutrinos are among the most abundant particles in the universe, nearly as common as the first photons.

Another major source of neutrinos is the sun. When researchers first began monitoring solar neutrinos in the 1960s, the number was about one-third of what they expected. The "solar neutrino problem" persisted for decades. It turns out that neutrinos come in three varieties, and they can transform into one another in midflight.

Neutrinos are so elusive because, with no charge and so little mass, the only way they can feel atoms is through the weak force, which comes into play only at short ranges. To notice an atom, a neutrino has to pass within about a proton's width of the nucleus. The nucleus is very tiny, so that doesn't happen very often.

ON THE FRONTIER

Because neutrinos penetrate matter with ease, they give researchers a unique view of corners of the universe that are otherwise hidden. The IceCube neutrino telescope, under construction near the South Pole, consists of thousands of light-sensitive detectors buried beneath the ice. High-energy neutrinos from beyond the solar system will sometimes strike an ice atom, emitting a flash of light for the detectors to spot. IceCube's goal is to make a neutrino map of the sky, which may give researchers a better understanding of extreme environments such as supernovas, which are believed to release 99 percent of their energy in the form of neutrinos.

COCKTAIL PARTY TIDBITS

- Nuclear physicist Enrico Fermi coined the name *neutrino*, which means "little neutral one" in Italian. Physicist Wolfgang Pauli predicted its existence in 1931 to explain why radioactive decay seemed to consume more energy than it produced.
- Each of our bodies emits about 400 neutrinos per second from decay of radioactive potassium in our bones. Fifty billion of them pass through each of us every second from radioactive elements in the ground.

DARK MATTER

THE BASICS

Way back in the first section we mentioned that most of the matter in the universe is not the ordinary stuff of which atoms are made. In fact, it's not any kind of matter that we currently know. We call it dark matter, because all we do know is that it must not absorb or reflect light. Measurements of microwave radiation left over from the Big Bang indicate dark matter makes up 23 percent of the universe. Only 4 percent of the universe consists of visible or "baryonic" matter.

If we can't see dark matter, how do we know it's there? Well, something is making galaxies behave oddly. Spiral galaxies such as the Milky Way rotate faster toward the center than toward the ends of their spiral arms, because matter gets thinner out there. But the difference in rotation speed is less than what you would expect, as if the galaxy were embedded in some additional stuff that made it more viscous.

As far as astronomers can tell, galaxies and clusters of galaxies are always sitting in blobs of dark matter, like Christmas lights decorating a tree. They can tell this by looking for gravitational lensing around galaxies, the focusing of light by dense clumps of matter. The best evidence yet for dark matter comes from the Bullet Cluster, a pair of colliding galaxy clusters. The gravitational lensing of the Cluster is strongest in two clumps that are set apart from the visible matter.

ON THE FRONTIER

Researchers think dark matter must consist of a blizzard of particles that feel only the weak force like the neutrino but are even less prone to bumping into atomic nuclei. They call these hypothetical oddballs weakly interacting massive particles, or WIMPs. (Physicists are pretty good with acronyms.) One candidate WIMP (there could be more than one) is the sneutrino, a neutrinolike particle predicted by supersymmetry, a proposed extension to the Standard Model.

Experiments are under way that could potentially detect WIMPs. And if the sneutrino exists, the Large Hadron Collider near Geneva has a chance to detect it in coming years.

COCKTAIL PARTY TIDBITS

- **WIMPs, should they exist, would be zipping through Earth at a rate of a few per liter per second. (A liter is the volume of a mid-size Coke bottle. Drink up.)**
- **If WIMPs exist, they could annihilate each other in the core of the galaxy and release detectable gamma rays.**
- **Some scientists have proposed that what we call dark matter is actually a change in the force of gravity across large distances. But this idea has a hard time explaining the lensing around galaxies as well as variations in the cosmic microwave background.**

VACUUM ENERGY

THE BASICS

Quantum mechanics tells us there's no such thing as empty space. Even if you could suck all the subatomic particles and photons out of a volume of space, there would still be fields such as the electromagnetic field and others we talk about later. According to the uncertainty principle, nothing can be completely still. No matter how much energy you sucked out of a field, if you looked closely at it you would find fluctuating pockets of energy happening everywhere.

Normally you would have to be a subatomic particle to notice these vacuum fluctuations, because they average out over distances much bigger than particles. But remember that according to general relativity, energy bends spacetime. This energy of empty space is called vacuum energy or the cosmological constant, and it has a surprising effect. Normal mass and energy exert pressure—they want to expand—but gravity forces them together.

Vacuum energy works the opposite way. It has what's called negative pressure, meaning it resists expansion, sort of like Silly Putty. And in general relativity, gravity causes anything with negative pressure to expand outward. And unlike matter, vacuum energy can't be diluted. It's spread evenly throughout space. So the bigger the universe gets and the more dilute regular matter becomes, the more vacuum energy will tend to push everything apart. This becomes very important later when we talk about the expansion of the universe.

ON THE FRONTIER

For now, let's note that vacuum fluctuations can actually be tapped in the lab. In the 1960s, Dutch physicist Hendrik Casimir figured out that if you brought two electrically neutral metal surfaces within a few thousandths of a millimeter of each other in a vacuum, some of the fluctuations in the electromagnetic field between them would cancel out. The fluctuations outside the plates would be unchanged, and they would in effect squeeze the plates together, as if the air pressure around the plates had increased.

COCKTAIL PARTY TIDBITS

- **The Casimir effect is slight: when surfaces are 10 nanometers apart, or about the size of 100 atoms, it generates a force equivalent to one atmosphere of pressure.**
- **There are people who claim to have tapped vacuum energy as a source of power, but you shouldn't believe them. A slight attraction between metal plates is hardly going to solve the world's energy problems.**

CHAPTER

EIGHT

THE MILKY WAY AND BEYOND

THE MILKY WAY

THE BASICS

The Milky Way is our home galaxy, a group of some 200 billion stars orbiting a common center. Just about everything you can see in the sky without a telescope is part of the Milky Way. It's that vast stretch of glowing material visible at night if you're outside a city. All the ancients knew for sure was it looked like spilled milk. The word *galaxy* comes from the ancient Greek word "*galaxias*," meaning milky.

The first person to detect the true nature of the Milky Way was Galileo, who in 1610 looked through his telescope and confirmed that the glow was actually stars blurred together. Today we know the Milky Way is a type of spiral galaxy, consisting of a disk of stars and gas measuring some 100,000 light-years across, with spiral arms extending like a pinwheel from the galactic center. The Milky Way looks like a glowing blur to us because we see it edge on, with all the stars packed together into a disk.

We reside on one of the smaller spiral arms, called Orion. The stars in the spiral arms are younger and brighter than those in the galactic center, because out here new stars continue to form from clouds of gas. The oldest stars in the Milky Way were born 12 to 13 billion years ago, but researchers believe the galaxy itself is even older, formed out of the primordial gas of the early universe.

ON THE FRONTIER

At the very core of the galaxy, centered on the galactic bulge, lies an enormous black hole, 3 or 4 million times as massive as the sun and extending roughly 10 million miles across. Called Sagittarius A* ("A-star"), researchers detected it by studying the orbits of surrounding stars, which were moving so fast they could only be orbiting something extremely massive. In 2008, astronomers captured a radio image of Sagittarius A* that had a resolution down to the width of a baseball on the moon as seen from Earth.

COCKTAIL PARTY TIDBITS

- The sun lies an estimated 26,000 light-years from the galactic center, orbiting it at a speed of 140 miles per second. It should complete a full orbit in 226 million years.
- Scientists estimate that up to 60 percent of stars in the Milky Way are binary systems, two stars orbiting each other, or multiple systems of three or more stars.
- Out in the galactic plane, where the sun resides, the average distance between stars is a few light-years. Our nearest neighbor star, Proxima Centauri, is a red dwarf 4.2 light-years away. *Proxima* means "nearest to" in Latin.

EXOPLANETS

THE BASICS

The possibility of alien worlds has fired scientists' imaginations for centuries. If life exists elsewhere in the universe, it will surely be found on a distant planet orbiting another star. In 1995, scientists discovered the first extrasolar planet or exoplanet orbiting a sunlike star: 51 Pegasi, 50 light-years away. As of December 2008, researchers had found 329 likely exoplanets, all within our galaxy, of course. In 2007 alone, 61 new planets were detected.

Researchers believe that 10 percent or more of stars may have planets. Most of the exoplanets discovered so far are enormous gas giants, up to dozens of times more massive than Jupiter but otherwise thought to be similar to our solar system's largest planet. Many of them, dubbed "hot Jupiters," orbit close to their parent star, which cooks their atmospheres to high temperatures.

The most common way of detecting an exoplanet is to look for the wobble of a distant star as an orbiting planet tugs it back and forth. The wobble shows up as periodic shifts in the color of the starlight. Another strategy is to look for the dimming of starlight as the planet passes or transits in front of its star. Both detection methods favor the discovery of massive planets far larger than Earth.

ON THE FRONTIER

Up until November 2008, astronomers had never convincingly imaged an exoplanet directly. That month, researchers imaged a gas giant estimated at three times Jupiter's mass orbiting the star Fomalhaut, 25 light-years away. It appeared as a pinpoint of light in images from the Hubble Space Telescope. A second team independently spied three gas giants orbiting the star HR 8799 at many times the Earth–sun distance. Their masses ranged from seven to ten Jupiters.

COCKTAIL PARTY TIDBITS

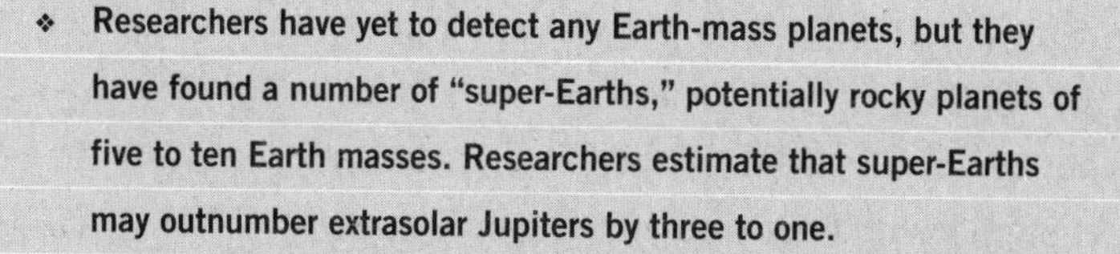

- **Researchers have yet to detect any Earth-mass planets, but they have found a number of "super-Earths," potentially rocky planets of five to ten Earth masses. Researchers estimate that super-Earths may outnumber extrasolar Jupiters by three to one.**
- **The most habitable exoplanet yet discovered is Gliese 581 d, the third planet out from the red dwarf star Gliese 581, about 20 light-years from Earth.**
- **In 2007, astronomers detected a record-setting fifth planet around the sunlike star 55 Cancri, 40 light-years away in the constellation Cancer.**

STAR CLUSTERS

THE BASICS

Millions of stars in the Milky Way come in clusters, groups of stars that were born together and remain bound together by gravity. Researchers study these clusters to test their theories of star formation and evolution.

Open clusters are loose groups of up to a few hundred stars, stretching up to 30 light-years across, often containing a lot of hot, young, blue stars. Born from the same clouds of gas and dust in the spiral arms, open clusters may retain some of the original gas, which keeps generating stars. The result is a mix of star types. Astronomers have discovered 1,100 open clusters in the galaxy, representing perhaps 1 percent of all its stars. Open clusters seem to be easily disrupted, losing their stars to the galactic center.

Globular clusters can contain as many stars as a small galaxy (10,000 to several million) packed into a dense sphere measuring 200 light-years across or less. They consist of ancient yellow and red stars, plus the occasional "blue straggler," a younger blue star thought to have formed from the merger of smaller stars. More tightly bound than open clusters, they are more stable but tend not to form new stars. Astronomers have found 150 globular clusters in the Milky Way, distributed roughly in a sphere around the galactic center.

ON THE FRONTIER

In 2005, astronomers discovered a third type of star cluster in the Andromeda Galaxy, our nearest neighbor spiral galaxy 2.5 million light-years away. Similar to globular clusters in composition, these intermediate clusters contain hundreds of thousand of stars but are spread across much larger areas of several hundred light-years, making them much less dense than the globular kind. The Milky Way doesn't seem to have any of these clusters, which fill a gap between globular clusters and "dwarf" galaxies.

COCKTAIL PARTY TIDBITS

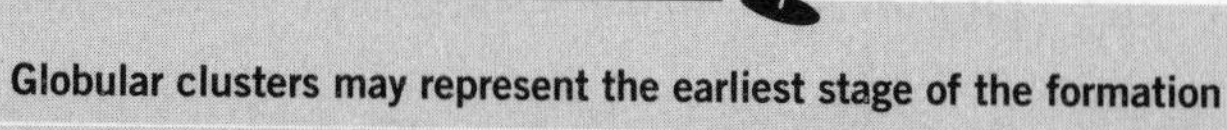

- Globular clusters may represent the earliest stage of the formation of the Milky Way, possibly predating the galactic disk.
- The most distant star cluster ever observed, containing hundreds of thousands of stars, is located about one billion light-years away.
- Once an open cluster has broken apart, the formerly bound stars will keep moving through space on similar paths. Such a group is called a stellar "association," or a moving group. Most of the stars in the Big Dipper were once part of an open cluster.

GALAXIES

THE BASICS

Not until the 1920s did astronomers confirm that spiral-shaped nebulas were actually distant galaxies like our own. Today, cataloguing galaxies is a minor industry. Modern galaxy surveys indicate the universe contains more than 400 billion galaxies, ranging in size from dwarf galaxies, one-hundredth the size of the Milky Way, to giant elliptical galaxies, spanning hundreds of thousands of light-years. Each one is home to anywhere from a few million to a trillion stars.

Galaxies come in three basic shapes: spiral, elliptical, and irregular. Spiral galaxies make up almost a third of nearby galaxies. They are like growing cities. The core of the galaxy—downtown, if you will—consists of millions of older red and orange stars packed densely into a sphere, called the galactic bulge. The spiral arms are like the suburbs where all the new construction is taking place. The stars here are younger and brighter, and new stars continue to form from clouds of gas.

Most other galaxies are elliptical, or lozenge-shaped. They tend to be full of older stars. Researchers believe they are the product of the collision and merger of spiral galaxies. Giant ellipticals are often found near the centers of big clusters of galaxies. Irregular galaxies, also known as "peculiar," may be in the first stages of merging with another galaxy.

ON THE FRONTIER

Most astronomers think large galaxies have so-called supermassive black holes at the center. Researchers aren't sure how these giant black holes form. They might come from regular-size black holes that slowly absorb vast quantities of matter; stars so big that they collapse directly into a huge black hole without producing a supernova; dense clusters of stars; or they may be remnants of the extreme pressures in the first moments after the Big Bang. The gigantic black hole's mass is proportional to the mass of the galaxy, which suggests that it plays a role in regulating the size of the galaxy.

COCKTAIL PARTY TIDBITS

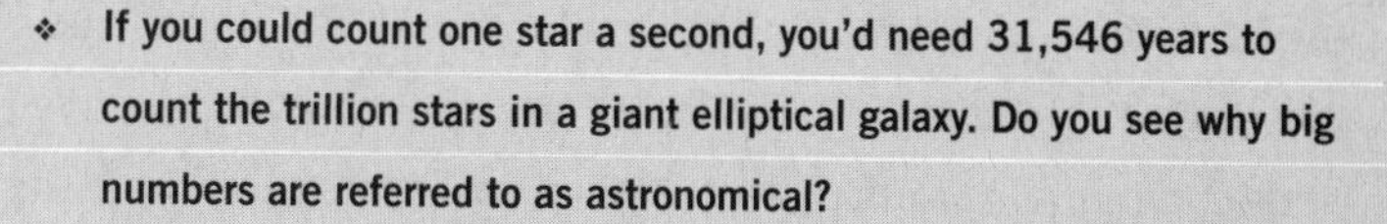

- **If you could count one star a second, you'd need 31,546 years to count the trillion stars in a giant elliptical galaxy. Do you see why big numbers are referred to as astronomical?**
- **Before about eight billion years ago, bigger galaxies were the ones making all the stars. Since then, smaller galaxies have taken up the slack. Researchers call this switch downsizing, and it may have had something to do with the growth of supermassive black holes.**
- **Some galaxies are lenticular, or lens-shaped, which could be spirals that have lost most of their outer gas and dust.**

ACTIVE GALAXIES

THE BASICS

Some of the best evidence for black holes in large galaxies comes from the existence of active galaxies, whose cores emit much more energy than would be expected from their component stars, gas, and dust. This energy can range across the electromagnetic spectrum, from radio waves to gamma rays. The most famous active galaxies are quasars, short for quasi-stellar objects, which shine brightly despite being up to 12 billion light-years away.

Researchers have long believed that active galaxies get their energy from supermassive black holes that have gathered a large accretion disk. As matter falls into the black hole, its gravitational energy is released as heat and electromagnetic radiation. About 10 percent of AGNs (active galactic nuclei) produce a pair of ultra-high-speed or relativistic jets of charged particles pointing in opposite directions.

One leading idea holds that the differences between active galaxies largely amount to differences in which way they are pointing. When the relativistic jets of an active galaxy are pointing toward us, the result is a blazar, an active galaxy dimmer but more energetic than a quasar that varies in output over minutes to days. Seyfert galaxies, which are less energetic than quasars but closer to Earth, may be quasars that face us edge on, such that the gas and dust around the central black hole blocks some of its light.

ON THE FRONTIER

Until the mid-1990s, astronomers thought active galaxies were an early-universe kind of thing, more common back when galaxies were bumping into each other left and right. The number of quasars seemed to peak around eleven billion years ago. Then they discovered a previously unknown group of quasars, hidden by gas and dust, that were active about eight billion years ago. Weaker individually than earlier quasars, their combined glow outshone their ancient counterparts, indicating that black holes were busy munching up galactic material much later than researchers had believed.

COCKTAIL PARTY TIDBITS

- **In 1918, astronomers observed a jet 5,000 light-years across streaming from active galaxy M87, a giant elliptical galaxy 60 light-years away in the Virgo cluster.**
- **Relativistic jets can appear to exceed the speed of light if they are pointing almost straight toward Earth but slightly off center.**
- **Up until the late 1980s, astronomers thought quasars might be white holes, inside-out black holes that spewed out all the energy the black hole captured.**

THE LOCAL GROUP AND GALAXY CLUSTERS

THE BASICS

Galaxies are by no means the largest structures in the universe. Galaxies congregate together in groups, clusters, and larger structures called superclusters. The spacing between galaxies is typically in the millions of light-years. But remember, gravity knows no bounds. It reaches out and holds galaxies in mutual orbits just as stars do in galaxies and planets around stars.

A so-called compact group is small and isolated, typically containing fewer than 50 galaxies. Our own Milky Way galaxy is one of about 40 members of the so-called Local Group, which extends for six million light-years. Other members include the Andromeda galaxy, 2.6 million light-years away, and the Large Magellanic Cloud, a satellite galaxy of the Milky Way 169,000 light-years away. The Local Group is expected to remain intact for trillions of years.

A galaxy cluster may have up to 1,000 galaxies, but they are typically packed into the same volume as a group, so the difference is mainly one of density. Beyond the Local Group is the Virgo cluster, a denser collection of hundreds of galaxies all jumbled together. The Coma cluster is even denser and has a neat spherical shape, centered on several giant elliptical galaxies. Like our Local Group, these galaxy clusters are a few million light-years in size.

ON THE FRONTIER

Scientists estimate that the Milky Way will collide with the Andromeda galaxy in as little as two billion years, merging to form a single galaxy, "Milkomeda." Andromeda is currently 2.3 million light-years from our galaxy. Researchers know that the two neighbors are approaching each other at 120 kilometers per second. Andromeda might be moving fast enough to the side to miss us, but if not, the two galaxies would pass through each other once or twice as they settled into one galaxy. This close encounter could easily kick our solar system to the farthest reaches of the galaxy, or even land us in Andromeda.

COCKTAIL PARTY TIDBITS

- **An isolated galaxy is called a field galaxy. About 5 percent of galaxies found in surveys fit this type. They can support more star formation than other galaxies because they have not lost their gas to interactions with other galaxies.**
- **Many components of galaxy clusters may be essentially invisible, including faint dwarf galaxies too diffuse to observe clearly and "dark" galaxies stretched too thin to make stars.**

GALAXY EVOLUTION

THE BASICS

Researchers would like to understand how galaxies merged into the forms we see today. But the images in telescopes are mere snapshots of the universe at different moments in time. It's like looking at aerial views of the major cities of the world taken once per century. To piece together the origins of galaxies, researchers keep building more powerful telescopes to peer farther away, which means further back in time.

Among local galaxies, only about one in a million appears to be interacting with another galaxy. More distant galaxies seem to experience collisions more frequently, indicating they were more common when the universe was younger. In fact, researchers think mergers were one of the driving forces behind the evolution of galaxies. According to the going theory, galaxies started off small and combined to make spiral galaxies. When spirals merged, they shed the gas in their disks and their stars were kicked into more complicated orbits. One common end result: an elliptical galaxy.

Supporting this idea, giant elliptical galaxies are found only in the middles of dense galaxy clusters, where collisions would seem to be likely. The collision theory also explains why elliptical galaxies tend to be full of older stars, whereas spiral galaxies are still experiencing star formation. The spirals have yet to run out of gas.

ON THE FRONTIER

NASA plans to launch the James Webb Space Telescope, the successor to the Hubble Space Telescope, by 2013. It should be powerful enough to detect the first galaxies. But it won't complete the whole picture of the early universe. The universe went through a long dark age before the first galaxies formed. To probe this era, researchers are building a number of powerful radio telescopes.

COCKTAIL PARTY TIDBITS

- When galaxies collide, they can trigger bursts of rapid star formation that may last for only about 10 million years. These starburst galaxies were more common when the universe was young, but researchers estimate they still account for some 15 percent of star formation.
- The Milky Way is a cannibal, gobbling up smaller dwarf galaxies that cross its path. One of these, a ribbon of stars called the Virgo Stellar Stream, covers 5 percent of the sky in the northern hemisphere in the direction of the constellation Virgo.
- Most galaxies in the universe seem to be dwarf galaxies, containing a few million to a few billion stars packed into a region of about 1,000 light-years. They may be the leftovers of the first galaxies.

LARGE-SCALE STRUCTURE

THE BASICS

Surveys of galaxies over the past few decades have found that just as galaxies are bunched into clusters, the clusters are linked into chains called superclusters, which may stretch for hundreds of millions of light-years. The superclusters are themselves linked into walls and sheets, although there don't seem to be clusters of superclusters.

Our Local Group of galaxies is part of the Virgo supercluster, a forked and twisting chain of galaxy clusters that connects us to the Virgo Cluster, 52 million light-years away, and stretches for more than 200 million light-years, eventually merging with other superclusters. Clusters and superclusters can exist because their mutual gravity holds them together even in the face of cosmic expansion.

The distribution of galaxies is like foam, with sheets or filaments of galaxies separated by massive voids. One of the most dramatic examples of this structure is the Great Wall, a vast web of galaxies more than 500 million light-years wide. Astronomers have mapped the universe to a distance of a few hundred million light-years, and as far as they can see, the foam pattern repeats itself in all directions instead of forming even bigger structures. This repetition is called the "end of greatness."

ON THE FRONTIER

The large-scale structure of the universe is an echo of the Big Bang. Gravity has had time to draw individual galaxies into clusters but researchers believe it could not have arranged them into superclusters. Instead the matter in the early universe was slightly rippled, and matter collected in filaments around those ripples as water does in a river delta. The densest groups of filaments merged into superclusters. Driving the process was the dark matter in which the filaments were embedded.

COCKTAIL PARTY TIDBITS

- The Virgo Supercluster is falling toward a hidden source of gravity called the "Great Attractor," most likely a big supercluster obscured by the Milky Way.
- The voids between superclusters may not be quite as empty as they seem. Researchers suspect they contain clouds of hydrogen gas that are detectable only by the way they interfere with light from distant quasars.

CHAPTER

THE BIG BANG

THE BASICS

Okay, we're finally ready to tackle the big kahuna, the universe itself. The study of the universe as a whole is called cosmology. A theory of cosmology takes everything we know about the large-scale features of the universe and ties it all together. The theory which does that is called the Big Bang. The Big Bang is to cosmology what evolution is to biology. It's the idea that makes everything hang together.

It says that approximately 13.7 billion years ago, all matter and energy in the observable universe—all 92 billion light-years of it—was concentrated in a single point. And all that matter and energy in the universe was compressed into that space. Remember the conservation of energy: it must have been there from the start. So the universe started out extremely hot and dense, but that matter and energy caused the fabric of spacetime itself to expand, gradually lowering the temperature and spreading everything out.

There are three key pieces of evidence for the Big Bang: the fact that distant galaxies are receding from us at high speed, the faint glow of radio- and microwaves all through space, and the abundance of hydrogen and helium in the universe. The Big Bang model also allows us to understand why galaxies are clumped together the way they are.

ON THE FRONTIER

Nobody knows what caused the Big Bang. Just as at the core of a black hole, at the earliest moment of the Big Bang, Einstein's equations of general relativity indicate that the universe was infinitely dense, a sign the theory needs refinement to accommodate that moment. Until we understand what was going on at the moment of the singularity, we won't have much chance of explaining why the Big Bang happened. If researchers can work out a theory that mixes quantum mechanics with gravity, that should resolve the problem. Until then, all we can do is trace the consequences of the Big Bang to the present day.

COCKTAIL PARTY TIDBITS

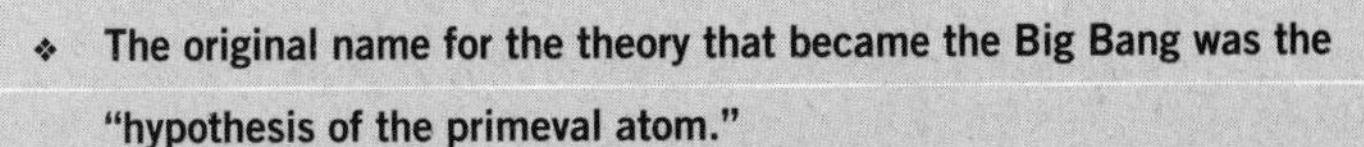

- **The original name for the theory that became the Big Bang was the "hypothesis of the primeval atom."**
- **Physicist Fred Hoyle coined the term "Big Bang," although he meant it derisively. Hoyle was a stubborn believer in the alternative Steady State cosmology, in which the universe had always been here and always would be.**

SPACE IS EXPANDING

THE BASICS

Early in the twentieth century astronomer Edwin Hubble discovered that of the dozen or so distant galaxies he could observe, they were all receding from us at great speed. Hmm, so if galaxies are getting farther apart, where are they all coming from? Hubble had uncovered the first piece of evidence for the Big Bang.

He succeeded in measuring the redshift of distant galaxies. Recall that for objects moving rapidly away from us, the wavelength of light coming from them will be stretched like a slinky. According to general relativity, space itself—the distance between two things—can also expand and contract. So the expansion of space shows up as a redshift.

An expanding universe is not like an explosion of galaxies into space. There's no center to the expansion, as there would be if a bomb exploded. It's an expansion of space itself. It's like a balloon with a ruler printed on it. As you inflate the balloon, the distance between ruler markings grows, and the farther apart the markings, the more rapidly that distance increases.

It's the same with galaxies. The more distant they are from us, the more they are redshifted. The expansion rate of the universe, the relationship between distance and redshift, is referred to as the Hubble constant. To our best understanding, it currently stands at 50,000 miles per hour for every million light-years between galaxies.

ON THE FRONTIER

The observable universe is any patch of space that we could ever receive a signal from, be it light, neutrinos, or gravitational waves. The distance in light-years to a far-off galaxy is not its current distance from us. The galaxy is long gone from that location, carried off by the expanding universe.

Taking into account the measured expansion rate, the observable universe is estimated to be a sphere measuring 92 billion light-years across. As the expansion rate of the universe changes (see the section on dark energy) the size of the observable universe will also change.

COCKTAIL PARTY TIDBITS

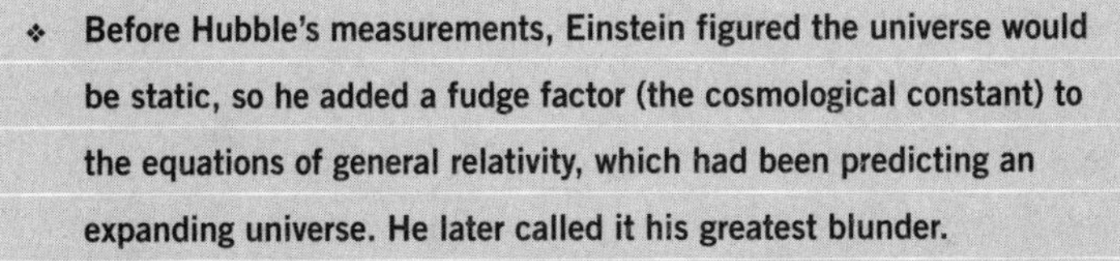

- **Before Hubble's measurements, Einstein figured the universe would be static, so he added a fudge factor (the cosmological constant) to the equations of general relativity, which had been predicting an expanding universe. He later called it his greatest blunder.**
- **There's no reason to think the edge of the observable universe is the edge of the universe itself. That would mean Earth is at the center of the universe, and scientists consider that unlikely.**

THE COSMIC MICROWAVE BACKGROUND

THE BASICS

One of our strongest pieces of evidence for the Big Bang is a faint glow of microwaves in the sky coming from no particular star or galaxy, called the cosmic microwave background (CMB) radiation. Ground- and space-based radio telescopes have confirmed that the CMB is highly uniform. Any theory of how the universe came into being has to explain this radiation and the pattern of its variations.

According to the Big Bang theory, the CMB is an impression left over from the first moment when protons and electrons condensed into hydrogen atoms 379,000 years ago. Prior to that point, the residual heat from the Big Bang prevented hydrogen from forming, and photons could travel only a short distance before being absorbed by the fog of electrons. The intensity of the CMB at different wavelengths corresponds to a temperature of about 2.7 kelvins, or nearly –459 degrees Fahrenheit.

The photons must have started out much hotter, but as space expanded, it stretched or redshifted the radiation, causing it to grow fainter as it must now fill a much larger volume of space. By taking into account the expansion of space over the past 13 billion years or so, they can estimate when the universe passed through that temperature and therefore how old the CMB is.

ON THE FRONTIER

The cosmic microwave background varies in intensity from place to place, because the soup of particles wasn't uniform. Some places were a little denser than average; other places were more diffuse. These density variations spread like the ripples when a stone is dropped into a pond, and the patterns in them tell us how much matter and energy the universe contains.

In 2003, high-precision data from the Wilkinson Microwave Anisotropy Probe (WMAP) told us the universe is 13.7 billion years old and is made of 4 percent normal matter (baryons), 23 percent dark matter, and 73 percent dark energy. The Planck Surveyor, a joint mission of NASA and the ESA, is slated to launch in spring 2009 to make finer measurements.

COCKTAIL PARTY TIDBITS

- **The variations in the CMB are pretty small. The microwaves are identical from one point in space to the next to at least one part in 10,000.**
- **Physicists Arno Penzias and Robert Wilson discovered the CMB in 1964 after years of trying to eliminate noise in their radio telescopes at Bell Labs that persisted despite eliminating all other signals.**
- **You can actually tune into the CMB on an old analog TV set. The CMB accounts for 3 percent of the static between channels.**

THE UNIVERSE IS FLAT

THE BASICS

Remember we said that spacetime bends under the weight of matter and energy. It has a shape. Well, the universe as a whole must also have a shape. Crucially, this shape must satisfy scientists' central assumption about the universe, that it looks the same in every direction. In spacetime terms, the curvature between any two galaxies in opposite corners of the universe should be the same.

There are only three basic shapes that fit the bill. To picture them, we have to resort to (imperfect) two-dimensional analogies. The first is like the surface of a sphere; we say it has positive curvature, meaning it bulges out in every direction. The second shape (negative curvature) shrinks inward in every direction, kind of like the surface of a saddle. The third shape is like a piece of paper. Called flat space, it has zero curvature.

The shape of the universe depends on the average density of matter and energy in it. By carefully measuring the cosmic microwave background, researchers have found that the density seems to be perched right on the knife-edge of flatness. A little more density and spacetime would bulge like a big ball; a little less and it would bow like a saddle. Called the critical density, it is roughly the mass of five hydrogen atoms per cubic meter.

ON THE FRONTIER

It would be nice to know if the universe were infinite—whether it goes on forever—but unfortunately, general relativity doesn't provide the answer. It only tells us how spacetime bends between any two points. But there is still a way to tell. If the universe were finite, it would act like a game of Pac-Man. Anything leaving the edge of the "screen," the universe, would emerge on the other side. Researchers can check for that possibility by looking for repeating patterns in the sky, like an image in a funhouse mirror. So far they haven't found any evidence of repetition. If the universe is finite, it must still be pretty big.

COCKTAIL PARTY TIDBITS

- **The flatness or near-flatness of the universe is now seen as an early sign that most of the universe is not made of visible matter. Surveys found only enough visible matter to account for 5 percent of the critical density.**
- **As recently as 2003, scientists speculated that the universe could be shaped like a dodecahedron, a twelve-sided object (think soccer ball), but the data now seems to rule out that possibility.**

INFLATION

THE BASICS

The flatness of the universe is hard to explain using the unadulterated Big Bang theory. If the universe started off just a hair above or below the critical density, then time and gravity would have amplified that mismatch by an enormous amount. Scientists don't like theories that have to be set up just so for things to work out right. It seems fishy.

They think the answer lies in the cosmic version of runaway inflation. In the first tiny fraction of a second of the Big Bang, the observable universe went from being many billionths the size of a proton to the size of somewhere between a marble and a football field. That's like blowing up a strand of DNA to the size of the Milky Way. Even the past 14 billion years of expansion hasn't enlarged the observable universe by that amount.

To go back to our analogy of an expanding balloon, if you blew up a balloon to the size of Earth, it would look flat instead of curved. So two points that started out very close together in the Big Bang would now be far apart. And any small deviation from flatness would be totally wiped out. It's the best explanation scientists have for why the universe is so close to flat.

ON THE FRONTIER

The details of the cosmic microwave background are consistent with inflation, but scientists wouldn't say it has yet passed the same kind of rigorous tests as the Big Bang. General relativity can accommodate an expanding universe. But to produce inflation, researchers have to add in an exotic form of energy, a type of vacuum energy or cosmological constant, that would have existed only for a brief moment during the Big Bang. Unfortunately, they don't have a theory to tell them exactly how much inflationary expansion would have occurred, which depends on how long the exotic energy stuck around.

COCKTAIL PARTY TIDBITS

- **Particle physicist Alan Guth, now a professor at the Massachusetts Institute of Technology, came up with the idea of inflation in late 1979. The United States had just gone through a period of economic inflation, which probably had something to do with the name.**
- **Inflation may depend on an exotic form of energy, but at least it doesn't require much of it. Researchers estimate that a mere 20 pounds' worth of energy would suffice to drive inflation.**

DARK ENERGY

THE BASICS

Researchers assumed for a long time that the expansion of space would eventually peter out under the combined tug of the gravity of all galaxies everywhere. If you throw a ball in the air, it comes down, after all. But in 1998, astronomers finished measurements of the redshift and distance of ancient supernovas and found that space had expanded more than they expected if galaxies were slowing one another down. In other words, the expansion rate of the universe is speeding up.

Researchers don't know what's causing the accelerating expansion, but according to general relativity, it must be some kind of energy, which—in homage to dark matter—they call dark energy. Measurements of the cosmic microwave background indicate dark energy must make up 74 percent of the universe, spread thinly and evenly throughout the universe.

The simplest form of energy would be something called a cosmological constant, which would remain the same over time. This is the energy of empty space we encountered before. To everybody's chagrin, simple estimates of vacuum energy predict an enormous amount that would have blown galaxies far and wide long ago. The observed dark energy has a mysteriously small value. Researchers have had a hard time understanding why that should be.

ON THE FRONTIER

It's possible that the cosmic acceleration could slow down over time. To find out, researchers need to measure the expansion rate of the universe very precisely. One proposed experiment is the Supernova Acceleration Probe, or SNAP, which would measure the distance and velocity of some 2,000 supernovas a year. A second technique is to conduct a huge, high-resolution survey of galaxies, looking for slight changes in their shape caused by light crossing through dark matter like a marble rolled over a warped floor. Such a map would tell researchers how dark energy had influenced the structure of the universe over time.

COCKTAIL PARTY TIDBITS

- **Researchers believe the universe started to accelerate about five billion years ago. If it had begun much earlier, galaxies would have never had time to form.**
- **Dark energy must be spread thinly and evenly throughout the universe, at a density of a few protons per cubic meter. All the dark energy in the solar system would add up to the mass of a small asteroid.**
- **Dark energy might make atoms ever so slightly larger than they would be otherwise. Assuming that it permeates the whole universe, it may counteract a little bit of the gravitational attraction between protons and neutrons in the nucleus.**

THE FIRST THREE MINUTES

THE BASICS

Three minutes isn't a lot of time. It's enough to microwave a bag of popcorn, take a quick shower, or listen to a song on the stereo, maybe two, if they're punk rock songs. But if you're the Big Bang, three minutes is all you need to fill the universe with matter.

With so much concentrated energy at its disposal, the early universe was one big particle accelerator. Particles and antiparticles of all types would have been popping into existence and annihilating each other. With each passing moment, the cauldron expanded and cooled, and the universe passed a series of checkpoints, restrictions on its ability to make different particles because there wasn't enough concentrated energy.

After a microsecond, the universe cooled enough that quarks (the particles that make up protons and neutrons) began to "freeze out," or join into protons and neutrons instead of annihilating themselves against antiquarks. The neutrons, being slightly more massive than protons and therefore less stable, decayed into protons until there was only one neutron for every seven protons. Together they formed nuclei of hydrogen and helium in a 9:1 ratio, as well as lithium.

Big Bang nucleosynthesis, as it's called, took about three minutes and created 98 percent of the helium in the universe today. (The rest comes from stars.) The ratio of hydrogen to helium and lithium is another pillar of evidence for the Big Bang.

ON THE FRONTIER

Just after the Big Bang, when the universe was very young and hot, it should have produced lots of particles and antiparticles, which would have annihilated each other. If antimatter were a perfect mirror image of matter, the universe would have made equal numbers of both, and the universe would be empty of matter. We're here, so obviously that didn't happen. This triumph of matter is dubbed baryogenesis ("genesis" for creation, and "baryo" for baryons, protons and neutrons).

COCKTAIL PARTY TIDBITS

- The title of this section is taken from the name of a famous book by particle physicist Steven Weinberg, who wrote that the more we understand about the universe, the less our place in it seems to matter.
- Astronomers measure the abundance of the primordial deuterium—heavy hydrogen—by studying light from distant quasars shining through invisible hydrogen clouds.
- Scientists think that during the baryogenesis, the universe produced about a billion and one regular particles for every billion antiparticles.

CHAPTER

TEN

PARTICLE PHYSICS AND UNIFICATION

THE STANDARD MODEL

THE BASICS

The Standard Model of particle physics is a pretty boring name for a hugely important theory. Over the past 30 years, researchers have pieced together the detailed workings of the three major forces at play inside atoms: electromagnetism, which holds atoms together; the strong interaction, which holds the nucleus together; and the weak force, which governs radioactive decay.

The Standard Model says the basic ingredients of the universe are not particles but fields, analogous to electric and magnetic fields, spread across all of spacetime. Every type of elementary particle is a unit of a different field. Even when no particles are around, these fields are constantly fluctuating, due to the uncertainty principle. When particles smash together, the interplay of the fields determines which particles decay and which new particles are born.

In the first few moments after the Big Bang, the universe was essentially one big particle accelerator, a cauldron of energy from which particles and antiparticles were popping up left, right, and center. The Standard Model tells us which particles were present at each moment, which ones survived, and how they settled into atoms. Along with general relativity, Einstein's theory of gravity, the Standard Model tells us how the universe came to appear the way it does today.

ON THE FRONTIER

Or close to it, at least. The Standard Model leaves a few questions unanswered, such as the nature of dark matter, and why the universe apparently has a slight preference for matter over antimatter. Answers to some of these questions may entail only modest changes to the theory, but others may lead to an even deeper theory. The most direct way to test the Standard Model is to build bigger and bigger particle accelerators, machines for smashing particles together at high energy. But these can get pretty expensive, so scientists are always looking for clues from astrophysics, such as neutrinos and cosmic rays.

COCKTAIL PARTY TIDBITS

- The Standard Model is the most accurate and precise theory in all of science. It correctly predicts the strength of the magnetic field around an electron to an accuracy of 1 part in 10 billion.
- The Standard Model has 20-odd "free parameters," constants of nature such as the charge of the electron and the masses of the quarks that scientists can't explain from first principles. They can measure the constants only in experiments.

PARTICLE ACCELERATORS

THE BASICS

One of our best tools for understanding the inner workings of the universe is the particle accelerator, a machine that whips protons, electrons, or other particles to high speed and crashes them together. From the collisions are born new particles that didn't exist before, which decay, or break down very rapidly, into yet other particles.

To get an idea why, remember the uncertainty principle. As we discovered with vacuum energy, even when a field is empty, it's still full of activity. Particle physicists find it handy to view the field as full of things called virtual particles bubbling up from the vacuum in pairs, one particle and its antiparticle.

The pair normally annihilates itself so quickly as to be undetectable, and these particles obey other funny rules, so they are called virtual particles to distinguish them from the everyday kind. But if you squeeze a lot of energy into a small space, you can give virtual particles wings, as it were, and promote them to the real world.

The particles you can make depend on how much energy you squish into a given space. The more concentrated the energy, the more massive the particles you can make, courtesy of $E=mc^2$. So a particle accelerator is essentially a giant microscope, but instead of magnifying small distances it squeezes energy into them.

ON THE FRONTIER

To comb those high energies—and short distances—requires building bigger machines to accelerate particles ever closer to the speed of light. You've probably heard about the Large Hadron Collider, a circular accelerator 17 miles around that uses powerful magnetic fields to twist beams of particles around a racetracklike path and smash them together.

The LHC is the biggest science experiment ever built. It is designed to whip opposing beams of protons to 99.9 percent the speed of light and collide them at the highest energies yet achieved by researchers.

COCKTAIL PARTY TIDBITS

- Scientists have compared particle accelerator experiments to studying how a television works by dropping it from the Empire State Building and combing the debris.
- The energy of subatomic particles is measured in electron volts. An electron volt is the energy imparted to a single electron by a one-volt battery.
- A few, shall we say, concerned citizens fear the LHC particle collisions might unleash world-devouring micro–black holes. Scientists have pointed out that cosmic rays regularly reach the same energies, and yet tiny black holes do not rain down upon us.

PASS THE BOSONS

THE BASICS

According to the Standard Model, particle physics is like a big game of dodgeball. The theory divides particles into two families, fermions and bosons. Fermions are the matter particles, the quarks and leptons. They behave as electrons in that they resist being squished together. They are like the dodgeball players. Bosons can all pile up in the same spot. They are responsible for transmitting the forces, like dodgeballs thrown between fermions.

The most famous force-carrying boson is the photon. In the Standard Model, when two electrons are put close together so their electric fields repel each other, what's happening is they are tossing photons back and forth. (Technically, these are undetectable "virtual" photons.) When an electron catches a photon, it wants to run off in a new direction. Hmm, so maybe it's more like laser tag than dodgeball.

Either way, the same thing goes for the other forces in the Standard Model. The strong force is transmitted by particles called gluons (a play on "glue"). The weak force operates through the W boson, which can be positively or negatively charged, and the neutral Z boson. The effective range of a force depends on the mass of its bosons. The weak force fades quickly because its bosons are massive. For gluons, which have no mass, it's more complicated.

ON THE FRONTIER

Why did the universe divide particles into bosons and fermions? Another good question! The Standard Model doesn't explain why. Here's where an idea called supersymmetry may help. It says that for every known boson, there's a heavier fermion, and for every known fermion, there's a heavier boson. These additional particles, called superpartners, have quirky names such as the selectron, photino, and squark. One hope is that the Large Hadron Collider, the new particle accelerator near Geneva, would spot some of these particles. If so, then the question changes: instead of why bosons and fermions, it's why are some bosons and fermions different from the others?

COCKTAIL PARTY TIDBITS

- **Bosons are named for Indian physicist Satyendra Nath Bose, who developed the theory of bosons along with Albert Einstein. Fermions are named for Enrico Fermi.**
- **Quantum principles imply that gravity ought to have its own boson, the graviton, which would have no mass and an infinite range, like the photon.**
- **By concentrating energy, particle collisions can promote a virtual particle such as a W boson into a real particle. That's how scientists proved the existence of the W and Z bosons once and for all.**

QUARKS

THE BASICS

If you could look inside a proton or neutron, you'd see each one actually contained three separate particles jiggling around inside it. These particles are called quarks. They are sensitive to all three forces of the Standard Model, but the strong force is what holds them together.

Researchers came up with the idea of quarks to explain the bewildering array of new particles that turned up when they began to smash atoms together in the 1950s. They proposed this particle zoo consisted of different combinations of two particles, dubbed the "up" and "down" quarks.

The proton is made of two up quarks and a down quark, whereas the neutron consists of two downs and an up. The up and down quarks have slightly different masses—the up quark is a little lighter—which is why protons are lighter than neutrons.

To make things more confusing, quarks also come in three colors, which are not in fact colors but more like electric charges. When quarks of the right colors meet, the strong force kicks in just as electricity does when positive charge meets negative charge.

Experiments have turned up four additional types or "flavors" of quark, like heavier versions of the up and down quarks. Only up and down quarks are stable in the everyday world. It takes high-energy collisions to produce the rest: strange and charm, top and bottom.

ON THE FRONTIER

The top quark is the heaviest known particle by far. Discovered in 1997 in experiments at Fermilab in Batavia, Illinois, it has a mass-energy of 180 GeV. That's as much as a whole atom of gold! Scientists still aren't sure why it's so big. They hope they can figure that out if they discover the Higgs boson, the hypothetical particle that gives mass to other particles, at the Large Hadron Collider in coming years.

COCKTAIL PARTY TIDBITS

- **Particles made of three quarks are called baryons. (Hence ordinary matter is called baryonic matter.)**
- **Could quarks be made of smaller particles? There's no reason we know of why they should, but if they are, these particles would be called preons.**
- **The top quark is so unstable that the strong force doesn't even have time to act on it. Instead, the weak force kicks in right away and breaks it down into a bottom quark and a W boson. It may be the only way we'll ever study a naked quark.**

THE STRONG FORCE

THE BASICS

Quarks were hard to discover because you can't break a proton or neutron into its component quarks. Unlike the other forces, the strong force has the funny property that it gets stronger as the quarks get farther apart. So quarks are like escaped prisoners in the movies, running through the woods chained at the ankle. As long as they stay close together, they barely affect each other. But if they stray too far apart, the chain yanks them back together.

The strong force is transmitted by gluons, which come in eight types that are like mixtures of the three quark colors: red, blue, and green. Quarks and gluons are complicated, even by physics standards. Because gluons themselves have color, they exchange gluons with themselves. So there's a lot more going on inside a single proton or neutron than just three quarks sitting there peacefully.

In fact, the masses of the up and down quarks don't even add up to those of the proton and neutron. Instead, most of the mass in the nucleus comes from the kinetic energy of quark and gluons, which whir around like angry bees inside protons and neutrons. The kinetic energy imparts a mass (once again thanks to $E=mc^2$). All that buzzing accounts for 4 to 5 percent of the mass in the universe.

ON THE FRONTIER

Researchers can soften the edges of protons and neutrons by making quark–gluon plasma, as we noted when talking about the nucleus. When nuclei come together at high energy, the strong force starts to link up all the quarks present into a big cloud that quickly decays into other particles.

Scientists originally thought that when quarks slam together to make plasma, they should squirt out jets of particles in opposite directions. But instead what researchers see are single jets, like a bullet fired into water. This tells them the quark–gluon plasma is much denser than they expected, more like a liquid than a gas.

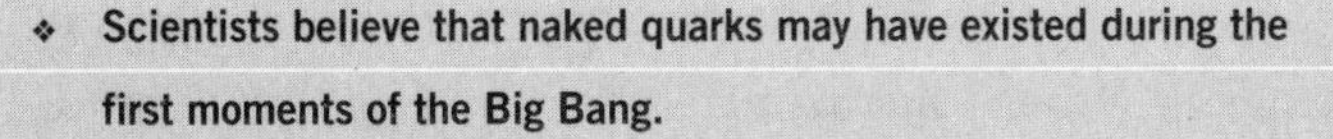

COCKTAIL PARTY TIDBITS

- Scientists believe that naked quarks may have existed during the first moments of the Big Bang.
- Some neutron stars may be dense enough to dissolve into lumps of quarks, called quark stars or strange stars. They would have only a thin crust of neutrons.
- One purpose of the Large Hadron Collider is to generate quark–gluon plasma by colliding lead ions, which should give scientists a fresh view of it.

MEET THE LEPTONS

THE BASICS

In the realm of matter, if you're not a quark, then you're a lepton, the category that includes electrons and neutrinos. Unlike quarks, electrons feel only electromagnetism and the weak force, and neutrinos are strictly creatures of the weak force.

They come in three pairs, called generations. The universe can't create one member of the pair without creating the other one, too. The most famous pair of leptons are the electron and the neutrino, which are both produced during radioactive beta decay.

The next pair also consists of a negatively charged particle, the muon, and a matching muon-neutrino. The muon is more massive than the electron but lighter than the quarks. It was the first particle discovered that isn't part of ordinary matter. It is also the most stable exotic particle. Muons rain down constantly on Earth, the product of cosmic rays exploding in the atmosphere.

The third lepton pair consists of the tau, also negatively charged but less stable than the muon, and the tau-neutrino. The tau particle exists only in particle accelerators. It has the mass of two protons, and when it decays, it creates particles made of quarks.

ON THE FRONTIER

All that's needed to make the Standard Model work are four particles: the up and down quarks, the electron, and the neutrino. But for some weird reason, like a person who has to have the same shirt in every color, the universe decided it needed two additional generations. Scientists think there must be only three generations or they would have found a fourth neutrino by now. Why three families and not four, or two? Nobody knows.

COCKTAIL PARTY TIDBITS

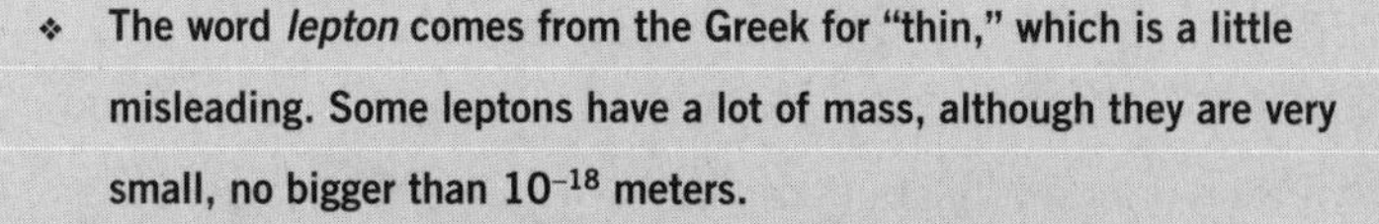

- The word *lepton* comes from the Greek for "thin," which is a little misleading. Some leptons have a lot of mass, although they are very small, no bigger than 10^{-18} meters.
- When the muon was discovered in 1936, it was so unexpected, theoretical physicist Isidor Rabi exclaimed, "Who ordered that?"
- Muons normally decay after 2.2 microseconds, but remember they can survive longer when moving at high speed as part of cosmic ray showers.

THE WEIRD FORCE

THE BASICS

The weak force ties together the four particles that make up ordinary matter in the universe: the up and down quark, the electron, and the neutrino. All four particles are capable of throwing around the W and Z bosons that transmit the weak force.

In radioactive beta decay, an up quark emits a W boson, which quickly breaks down into an electron and a neutrino. In the process, the up quark becomes a down quark, turning a neutron into a proton. Similar things happen in neutrino detectors and during nuclear fusion in stars.

Here's the weird part. The W boson weighs as much as 80 protons or neutrons. In dodgeball terms, it's as if your bouncy rubber ball transformed in your hands to a lead wrecking ball. The uncertainty principle allows for such strange behavior as long as the W can't get very far. That's why the range of the weak force is limited to inside the nucleus. The chances of a W making it any farther than that drop off sharply.

Still, the weirdness of the weak force cries out for an explanation, which researchers are pretty sure they have. It involves an exotic particle called the Higgs boson, which may show up in experiments over the next few years.

ON THE FRONTIER

The weak force is horribly biased. Electromagnetism treats all electric charges equally; ditto the strong force for color and gravity for mass and energy. But the weak force is always treating particles differently. For example, it's the only force that changes when researchers look at it in the mirror.

All fermions come in two varieties that are like mirror images of each other. One is left-handed; the other right-handed. The other forces don't distinguish between lefties and righties. But there are aspects of the weak force that only respond to left-handed neutrinos and right-handed antineutrinos.

Researchers are still investigating the matter, but right-handed neutrinos may not even exist. Ditto for left-handed antineutrinos.

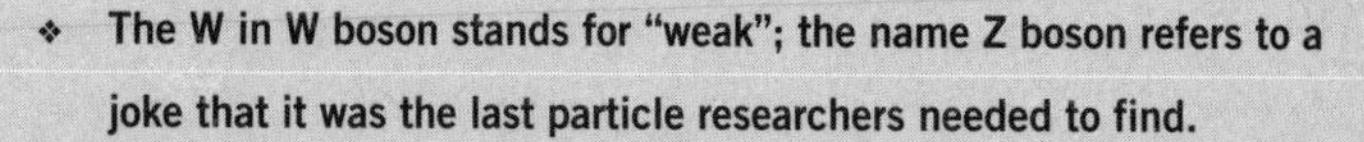

COCKTAIL PARTY TIDBITS

- **The W in W boson stands for "weak"; the name Z boson refers to a joke that it was the last particle researchers needed to find.**
- **Researchers predicted the existence of the W and Z particles in 1968. They finally detected the particles directly in particle accelerator experiments in 1983.**
- **The W boson is the only force particle capable of changing the generation of a particle, such as turning a strange quark into an up quark or a bottom into a charm.**

SYMMETRY

THE BASICS

Studies suggest that symmetric faces are easier on the eyes. The universe also seems to prefer symmetry. Researchers believe the rules of the Standard Model reflect different symmetries of nature. The motto of symmetry is, "The more things change, the more they stay the same." If you take a polished metal sphere and rotate it this way or that, your image in the sphere will not look any different. We say the sphere has rotational symmetry.

Similarly, experiments don't give different results from one moment to the next or in one corner of the lab versus another. Special relativity reflects the symmetry (called Lorentz symmetry) between experiments moving at different velocities, and general relativity extends the symmetry to accelerated motion. The symmetries of the Standard Model are much more abstract and complicated, but the idea is the same. The search for a theory that goes beyond the Standard Model involves looking for new symmetries.

Researchers believe that for many types of symmetry, there's something in the universe that is conserved. For example, symmetry of physical laws from moment to moment is connected to conservation of energy; rotational symmetry is connected to conservation of angular momentum.

ON THE FRONTIER

Some symmetries are broken, meaning the universe started out obeying it but something got in the way. Researchers know that antimatter is not in fact a perfect mirror image of matter. Sometimes when an antimatter particle decays or breaks down into other particles, it does so at a different rate than its matter counterpart. Another broken symmetry is thought to give rise to the mass of elementary particles.

COCKTAIL PARTY TIDBITS

- Sharp-tongued physicist Fritz Zwicky once said that a spherical bastard is a bastard any way you look at him.
- When we talked about left- and right-handed neutrinos, we were talking about something called CPT (charge/parity/time) symmetry, which is like turning your experiment into antimatter and looking at it upside down in a mirror.
- Symmetry is closely connected to the proposed unification of the three forces of the Standard Model at the high energies present during the Big Bang. As the universe cooled, symmetries were broken and the forces split into their current forms.

UNIFICATION

THE BASICS

One of the most ambitious goals of particle physics is to understand why the forces differ so widely in strength. Roughly speaking, gravity is millions of times weaker than electromagnetism, which is itself weaker than the weak force and weaker still than the strong force. The huge size range, known as the hierarchy problem, really bugs researchers because it seems too messy.

Part of their solution is to imagine that at high energies (or short distances) the forces start to become indistinguishable and ought to be unified into a single force, in the same way that electromagnetism fused electricity with magnetism and the electroweak theory added the weak force to the mix.

According to the Standard Model, the strengths of the forces change at shorter distances. Electromagnetism strengthens, whereas the strong and weak forces start to fade. Their strengths should become roughly equal at a distance of about 10^{-31} meters.

That's because particles are surrounded by a haze of virtual particles that shield their true properties. One effect of particle accelerators is to cut through vacuum fluctuations. As particles crack together with greater energy, more of the naked particle is exposed. The electron, for example, acquires a greater negative charge at higher energies, which effectively strengthens the electromagnetic force.

ON THE FRONTIER

Researchers have searched for a so-called grand unified theory (or GUT) that would bridge the strong and electroweak forces, but if there is such a theory, they haven't found it yet.

And even if they did find it, gravity would still be left out. Researchers believe it too grows stronger at shorter distances. They think it finally becomes the equal of the other forces at a distance of 10^{-35} meters, which is 10,000 times smaller than the domain of grand unification. They call that distance the Planck scale.

COCKTAIL PARTY TIDBITS

- **GUTs imply that the proton ought to be ever so slightly unstable. But if so, it must need more than 10^{-32} years to break down or experiments would have detected it by now.**
- **GUTs predict the existence of monopoles, individual magnetic poles analogous to electric charges.**
- **Supersymmetry would bring the forces into better alignment. But researchers know that if supersymmetry does exist, it must be broken or we would have found evidence of it by now.**

THE HIGGS BOSON

THE BASICS

The Higgs boson is the final missing ingredient of the Standard Model. It is thought to be responsible for giving other elementary particles their masses. Like other particles, the Higgs boson is a manifestation of an underlying field, called the Higgs field, naturally. But when other fields are drained of all their energy, their activity drops to a minimum. The Higgs field is different. At its lowest energy, it's still hanging around, doing something. And what it does is act like molasses.

When other elementary particles try to move around in the Higgs field, they set up vibrations in it, which slows them down. It's like a celebrity mobbed by a crowd. Some particles are A-list celebs and attract a bigger crowd (more mass); others are hanging out with screechy-voice comedian Kathy Griffin on the D-list.

The Higgs boson or something like it is needed to explain why electromagnetism and the weak force appear so different. In the 1970s, researchers discovered the two forces are part of a single electroweak force. According to this theory, the W and Z bosons should have no mass. But they do. The symmetry between the electro and weak parts of the force is broken. And the Higgs field is the scientists' leading idea for what does the breaking.

ON THE FRONTIER

The Higgs boson is the sexiest thing in particle physics right now, because it's like the cute neighbor next door—attainable! It may show up in experiments within a few years. Past searches and studies of the top quark have shown that if it does exist, the Higgs boson has to have a mass somewhere between 115 and 180 giga-electron volts. The Large Hadron Collider, the 17-mile-long circular particle accelerator near Geneva, is designed to generate energies up to 14,000 giga-electron volts, enough to make lots of Higgs bosons and study their behavior.

COCKTAIL PARTY TIDBITS

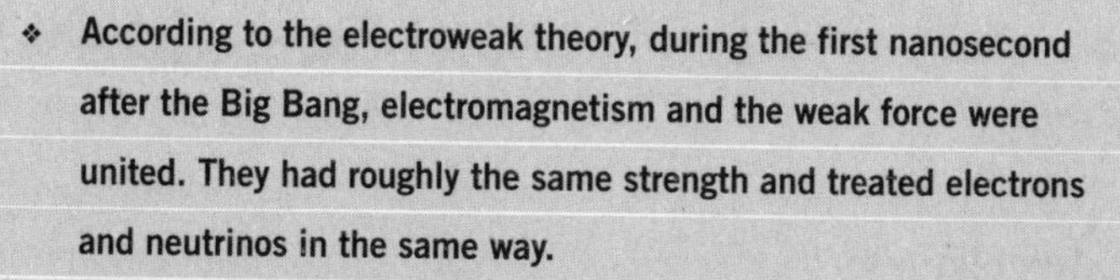

- According to the electroweak theory, during the first nanosecond after the Big Bang, electromagnetism and the weak force were united. They had roughly the same strength and treated electrons and neutrinos in the same way.
- Technically, the Higgs field is just a particular type of field, distinct from the matter fields and force fields in the Standard Model. If an idea called supersymmetry is correct, there should be additional Higgs fields with their own Higgs bosons.

QUANTUM GRAVITY

THE BASICS

General relativity covers the size and shape of the universe. It describes a smoothly curving spacetime. Quantum theory covers the behavior of matter and energy at the smallest scales and highest energies. Nothing quantum is completely smooth; everything becomes fuzzy when viewed up close.

Researchers would like to combine the two perspectives into a quantum theory of gravity, applying the lessons of quantum mechanics to the fabric of the universe. One thing they are pretty sure of is that according to quantum mechanics, spacetime should constantly fluctuate randomly like a flag in the wind. But when researchers apply their standard mathematical tricks, they find that the more you zoom in on spacetime, the more violently it churns until finally it's shredding itself apart.

Quantum theory implies that spacetime must be broken up into pieces, like foam. There should be a smallest possible distance between any two things. Known as the Planck length, in honor of Max Planck, the guy who kicked off quantum mechanics, it is 10^{-35} meters, or $1/10^{20}$ of the diameter of a proton. If you squish particles closer together than that distance, you need a quantum theory of gravity to understand what happens to them, and we don't have one of those yet.

ON THE FRONTIER

Researchers think a successful quantum theory of gravity will explain what happens at the singularity inside black holes and at the Big Bang, where general relativity says matter and energy become infinitely concentrated. Quantum gravity ought to enforce a cutoff, a limit to the density. If spacetime comes in pieces that can be only so small, then you can't squish the universe into a space smaller than that. The leading candidate for quantum gravity is string theory, in which nothing is smaller than strings; but there is another one called loop quantum gravity, in which there are "atoms" of spacetime that are smaller than strings.

COCKTAIL PARTY TIDBITS

- Space and time may not emerge unscathed from a quantum theory of gravity. Researchers suspect that when they find the theory, it will show that space and time don't exist as such at the Planck scale.
- String theory and loop quantum gravity indicate that there could have been a "big bounce," in which a prior universe collapsed on itself and expanded again to form our universe.

STRING THEORY

THE BASICS

String theory hypothetically unifies all of particle physics by showing that all the forces are made of the same underlying stuff. According to string theory, all particles are different manifestations of a single object called a string. Different vibrations of the same string correspond to different particles. It's like a guitar string: one note corresponds to the electron; another note produces the top quark; another is a photon, and so forth.

The potential for unification comes in because one of these vibrations has the properties of the graviton, the hypothetical particle that transmits the force of gravity. But to accommodate the different vibrations, strings need extra room to wriggle, in the form of six (or possibly seven) extra dimensions of space. In the everyday world, people and particles alike can only move either up–down, side to side, or forward–back. What string theory predicts is if you could shrink yourself by an incredible amount—far smaller than a proton in the nucleus, far smaller even than the quarks inside a proton—you would eventually reach a point where you could move in new directions. It's as if particles were like ants traveling on tiny clotheslines. From our perspective, the particle seems to be traveling in a straight line. But on closer inspection it's actually spiraling around the clothesline.

ON THE FRONTIER

The curled-up extra dimensions of string theory intertwine at every point in spacetime in a shape called a Calabi–Yau space. And therein lies the catch. (There's always a catch, right?) The masses and other properties of strings depend on the shape of the Calabi–Yau space. But string theory works equally well for an astronomical number of different possible Calabi–Yau spaces. Scientists could in principle narrow down the number, but first they have to figure out a way to observe obviously string-related effects, which so far they haven't been able to do.

COCKTAIL PARTY TIDBITS

- String theory is a bit of a misnomer, in that the term "theory" is normally reserved for something that has been tested and retested in numerous ways, such as the atomic theory of matter or the general theory of relativity.
- Critics of string theory say it's bad science because it's untestable, but its proponents say they are still working out the details. And nobody has come up with any better ideas.
- To get direct evidence of strings, we would need a particle accelerator the size of the solar system, and Congress probably wouldn't fund that.

CHAPTER

ELEVEN

THE OUTER LIMITS

ALIEN LIFE

THE BASICS

We know life exists on Earth. Scientists are as keen as anyone else on figuring out whether it exists elsewhere in the universe. Until recently, barring one of those alien invasion scenarios, all we could do was poke around for water in the solar system or listen for radio signals from extraterrestrial civilizations. But the discovery of planets around other stars has breathed new life into the search for aliens.

The odds of discovering life out there depend on how common extrasolar planets are, how frequently they might support life and for how long, and what sorts of signs those living things might generate. When looking for life in the solar system, NASA's mantra is "Follow the water." As far as we know, liquid water is the only stuff capable of supporting the complex chemical reactions that presumably gave rise to life on Earth.

The same goes for life outside the solar system. Scientists figure their best bet is to seek out Earthlike planets orbiting their stars at just the right distance so the planet can support liquid water on its surface. Too close and water vaporizes; too far away and it freezes. The range and extent of this habitable zone depend on the size and temperature of the star.

ON THE FRONTIER

The search goes on for Earthlike planets. With today's telescopes, planets of Earth's size are too small to pick out from their effects on stars. Future telescopes such as the James Webb Space Telescope, to be launched in 2013, have the potential to detect them and even observe them directly. The next step will be to analyze their atmospheres, looking for the kinds of changes that microscopic organisms can cause. Researchers think oxygen and methane (given off by microbes, cows, and other organisms here on Earth) might be the first things to check off the list.

COCKTAIL PARTY TIDBITS

- **Scientists might detect photosynthetic life on other planets, alien plant-life, by looking for absorption or reflection of characteristic colors. Plants around hot blue stars might absorb blue light, making them green, yellow, or red; those around dimmer red stars might try to absorb it all, yielding black plants.**
- **In 2007, researchers detected water vapor in the atmosphere of a gas giant planet by looking at the starlight shining through it. Unfortunately, such a planet wouldn't support pools of liquid water.**

ADVANCED CIVILIZATIONS

THE BASICS

Life on Earth is the only example we've got. Biologists can't say whether the kind of evolutionary pressures that gave rise to walking, talking, tool-wielding, road-building primates (that would be us, of course) would be common on other planets.

But to some scientists, it's only natural that intelligent life would have arisen many times by now in our own galaxy, and if so, those lifeforms should have had plenty of time to cross interstellar space. But the same scientists quickly realized that if such a race had emerged even once, the green buggers should be swarming all over the place.

Scientists think colonizing the galaxy would take somewhere between 5 and 50 million years, assuming that aliens traveled between stars at "slow" rates of 10 to 20 percent of the speed of light. All the aliens need do is plunk down a colony on the nearest habitable planet. As soon as the colony can build more rockets, the process repeats.

And yet we haven't spotted E.T. number one. This apparent contradiction is known as the Fermi paradox for nuclear physicist Enrico Fermi, who was very fond of estimating quirky things such as the time it would take aliens to colonize the galaxy.

ON THE FRONTIER

The Fermi paradox quickly leads to paranoid thoughts. Did the aliens destroy themselves? Are they shy? Or could it be they're already here, walking among us undetected?? As eggheads and not movie producers, we take a more rational view. Either we're not looking in the right places, which could be, or—what seems more likely—we're biased, and intelligent life is simply a lot less common than we assume. Or maybe a lot more fleeting; see the next chapter.

COCKTAIL PARTY TIDBITS

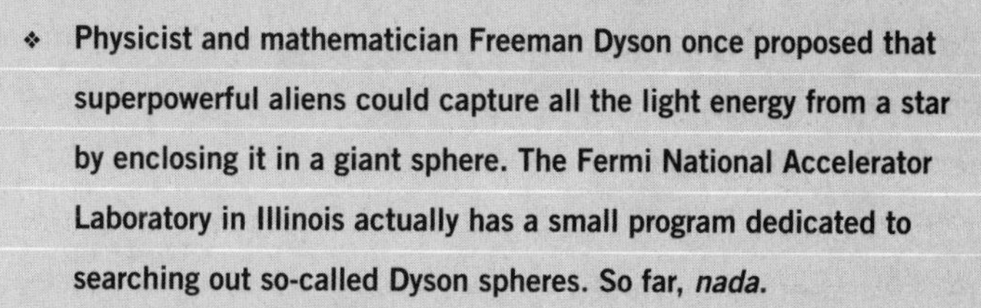

- Physicist and mathematician Freeman Dyson once proposed that superpowerful aliens could capture all the light energy from a star by enclosing it in a giant sphere. The Fermi National Accelerator Laboratory in Illinois actually has a small program dedicated to searching out so-called Dyson spheres. So far, *nada*.
- The SETI (Search for ExtraTerrestrial Intelligence) Institute in California listens for radio broadcasts from distant stars, but some scientists suspect that aliens might broadcast using neutrinos, which would reach a lot farther because they pass through matter.
- If aliens ever made direct contact, it would probably be through robots, according to SETI astronomer Seth Shostak, perhaps in the form of "von Neumann machines," self-replicating probes sent out to colonize the stars.

THREATS TO LIFE

THE BASICS

The universe is not a friendly place. Assuming we manage not to blow ourselves up, succumb to a killer flu strain, or change the climate faster than we can keep up, we're in the clear, right? Not necessarily. There are things in space that could close the curtains on us.

We've seen that collapsing stars produce enormous explosions called supernovas and gamma ray bursts. If either one of these events happened nearby in the galaxy, we would be showered by high-energy radiation that would fry Earth's ozone layer, which normally shields us from lethal solar rays. Without it, the x-rays and ultraviolet from a supernova would penetrate the atmosphere.

Astronomer Phil Plait estimates that being caught in the path of a gamma ray burst would be like detonating a one-megaton nuclear bomb over every square mile of the planet. Those of us who weren't incinerated would quickly succumb to radiation exposure as it burned our skin and zapped our internal organs, like the Japanese victims of the atomic bomb blasts in World War II.

Exactly how close these explosions would have to be are a little unclear, perhaps within 25 light-years for a supernova or 3,000 light-years for a gamma ray burst, but judging from our stellar neighbors, scientists don't think either one is very likely to happen.

ON THE FRONTIER

More likely would be an asteroid strike of the kind depicted in the movies *Armageddon* and *Deep Impact*. Plait has helpfully compiled the odds of a host of world-ending events. By his estimation, the odds of a supernova or gamma ray burst are 1 in 10 million and 14 million per century, respectively. An asteroid strike ranks much higher: 1 in 700,000.

Contrary to *Armageddon*, nuking an asteroid would probably not lead to a happy ending, as the giant rock would then be blown into a hail of smaller but still deadly missiles. Instead, we might nudge an asteroid by pushing on it with a rocket.

COCKTAIL PARTY TIDBITS

- **Barring unforeseen calamities, the fate of life on Earth rests with the sun. Plant life will die off in about 900 million years, as rising temperatures short-circuit the carbon dioxide cycle. Without plants to make oxygen, animal life will quickly go extinct.**
- **Scientists have proposed we might dodge the radiation of our sun's red giant stage. All we have to do is use asteroids to adjust our planet's orbit.**

VISITING A BLACK HOLE

THE BASICS

Black holes have a reputation for being among the most hostile places in the universe, and in many ways, it's deserved. Let's say you are falling toward a solar mass black hole head-first like a skydiver. You are well outside the event horizon, the distance at which nothing can escape a black hole. For a black hole as massive as the sun, that's two miles from the center.

But the high concentration of matter in the black hole generates an intense gravitational pull for hundreds of thousands of miles around, and the pull increases with every inch. By the time you got within about 400,000 miles of the black hole, the gravity would be so much stronger at your head than your feet that it would pull you apart in a long thin strand like taffy. This is called "spaghettification."

If you could survive that, say because you were really stretchy like Reed Richards of the Fantastic Four, you would see the stars around the black hole turn blue as you accelerated toward them; the stars behind you would appear red. When you passed through the event horizon, you wouldn't notice any change until a gravitational riptide caught you and sent you whirling around the singularity with the rest of the matter stuck inside the black hole.

ON THE FRONTIER

You wouldn't detect it as you crossed the event horizon, but scientists believe that black holes emit a faint glow. Recall that space is full of virtual particle–antiparticle pairs. In the 1970s, Stephen Hawking, the wheelchair-bound theoretical physicist, realized that for virtual pairs right on the horizon, one member of the pair might be pulled across while the other one got away.

Dubbed Hawking radiation, these particles should cause the black hole slowly to evaporate, drip by drip, by donating energy to the outgoing radiation. Emphasis on "slowly." A black hole the size of the sun would take around 10^{66} years to evaporate, or many times the current age of the universe.

COCKTAIL PARTY TIDBITS

- **If you could buzz the event horizon in a spaceship, staying far enough away to fly back out, you would find that thousands of years had passed when you returned to your mothership waiting well beyond the black hole. The intense gravity would have slowed time in your ship.**
- **In 2008, a team of researchers made an artificial event horizon by firing a pair of specially shaped laser pulses down an optical fiber, one after the other. The second, faster-moving pulse got stuck behind the first pulse, like light trapped behind an event horizon.**

FREE WILL VERSUS THE UNIVERSE

THE BASICS

We feel as if the things we do are completely spontaneous. Why are you reading this book? Well, because you want to learn more about the universe, obviously. And you're here on this page presumably because you thought this section sounded interesting. You probably feel as if you could have chosen to do something else if you had wanted to. This feeling is called free will, and it's one that's very important to us. Most of us would get depressed if we thought we had no control over our actions.

But according to the laws of physics, although there are many things that *could* happen, there is only one thing that *will* happen. We are made of atoms, and atoms obey rigid rules that determine exactly what will happen to them from one moment to the next. We say the universe is deterministic. So we feel as if our actions are determined on the fly. But on the other hand, everything that happens in the universe is determined by what came before. How do we reconcile the two views?

This is partly a question about how the mind works, which is beyond the scope of this book. (Check out *Instant Egghead Guide: The Mind.*) But from the standpoint of physical laws, we can observe that even very simple systems can show such complicated behavior to be essentially unpredictable.

ON THE FRONTIER

Wait a minute, you may be thinking. Didn't we learn earlier that in the quantum world events are nondeterministic, meaning they happen at random? That's true (although remember also that quantum wave functions are deterministic).

Mathematician Sir Roger Penrose has proposed that quantum effects in the brain could be much larger than we assume. Most scientists are pretty skeptical. Remember Schrodinger's cat. Once you start mixing particles together, quantum randomness quickly disappears.

COCKTAIL PARTY TIDBITS

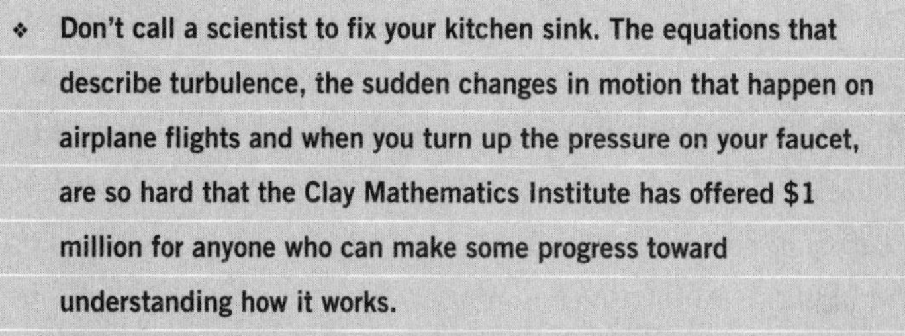

- **Don't call a scientist to fix your kitchen sink. The equations that describe turbulence, the sudden changes in motion that happen on airplane flights and when you turn up the pressure on your faucet, are so hard that the Clay Mathematics Institute has offered $1 million for anyone who can make some progress toward understanding how it works.**
- **When you hear somebody talk about the butterfly effect, they're referring to the idea that a slight change in starting conditions, say, atmospheric temperatures and pressures across the globe, can have a big effect on an outcome, like the weather.**

TIME TRAVEL AND WORMHOLES

THE BASICS

Speaking of the choices we make, there are probably a few in your life you wish you could take back, such as maybe investing all that money into the stock market a few years ago. Everybody wants to know, is time travel possible? Instant Egghead is here to tell you it's a definite, unequivocal, "We don't know."

General relativity doesn't outright prohibit travel backward in time. Researchers starting in Einstein's day were able to mathematically construct spacetimes in which time bent back on itself, but they always involved something that couldn't exist in our universe, like an infinitely long, rotating cylinder.

Perhaps the most plausible time travel scheme involves a wormhole, a hypothetical bridge directly connecting two points in spacetime, no matter how distant or separated in time they are. A wormhole is like a Web-cam connection, except you can reach through to the other side. Assuming you could find or make such a thing, you could put one end of the wormhole on your spaceship, fly really fast for a while so that a lot of time passed on Earth, and then come back.

You and your end of the wormhole would now be in the far future, but the other end of the wormhole would still be in the past.

ON THE FRONTIER

One caveat (among about a million) to wormhole-based time travel is that the spacetime-cam would collapse almost instantly without something to hold it open. Specifically, you'd need a source of so-called negative energy. About the closest anyone has come to negative energy is the Casimir effect, an attractive force. But by one estimate, the amount of energy you'd need to stabilize a wormhole one meter wide is almost as much as the sun would put out in 10 billion years.

COCKTAIL PARTY TIDBITS

- **Next time you're flipping through a physics journal, a trip backward in time in general relativity jargon is called a "closed timelike curve," and a wormhole is also known as a "multiply connected" spacetime.**
- **The idea of time travel leads to all sorts of fun paradoxes, such as whether you could go back in time and prevent your parents from meeting and conceiving you.**
- **There's also the paradox that if time travel is possible, why aren't we knee deep in twenty-fifth-century high school students studying our era for extra credit in history class? Maybe it's because a wormhole can take you back in time only as far as the invention of wormhole time machines. And that could be a loooong way off.**

MANY WORLDS

THE BASICS

If you can't go back in time to correct your ill-timed foray into the stock market, then wouldn't the next best thing be knowing that somewhere out there in a parallel universe lives a version of you that was a little bit wiser—or a little poorer—and didn't invest at all? Well then, here's some good news. Some scientists think that may actually be the case!

Recall that in quantum mechanics, a wave function eventually "collapses" into a definite state at random from among the many possible states into which it could have collapsed. In trying to come to grips with what that tells us about reality, a young physicist named Hugh Everett decided to bite the bullet. What if, Everett imagined, every time a wave function collapses, the universe splits off into an infinite number of parallel universes, one for each different outcome? In one universe, you invested in a different stock that didn't tank so badly when the economy took a tumble. In another universe, you were late for work, lost your job, and didn't have any money to invest.

Called the Many Worlds interpretation of quantum mechanics (because the movie called *Sliding Doors* hadn't come out yet), some smart people actually consider it a legitimate concept. Unfortunately, unlike other interpretations of quantum mechanics, there doesn't seem to be any way to test it. Then again, maybe we're better off not knowing.

ON THE FRONTIER

The Many Worlds idea at least makes thinking about time travel a lot more fun. Maybe when you go back in time, instead of being unable to change the past, you can make it split into a new timeline, Many Worlds–style. That's basically what was happening in *Back to the Future Part II*, when Michael J. Fox had to go into the future to fix problems he introduced by changing the past in the first movie. Got that?

Hey, so maybe there is a way to test Many Worlds. Just go back in time and see if you can change the past, although good luck convincing anybody what you'd done. They'd probably lock you up. But don't worry. In another universe, they made you their king!

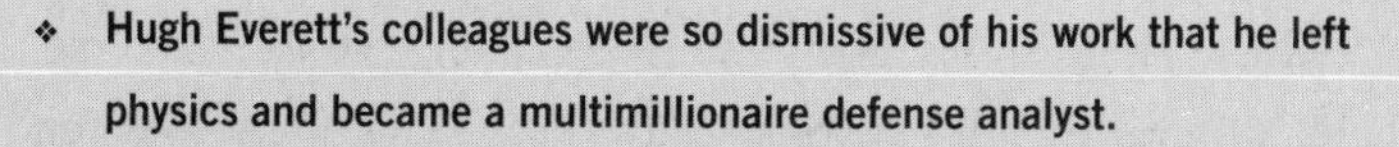

COCKTAIL PARTY TIDBITS

- **Hugh Everett's colleagues were so dismissive of his work that he left physics and became a multimillionaire defense analyst.**
- **Everett's son, Mark, fronted the mid-1990s rock band Eels and was the subject of an award-winning 2007 BBC documentary in which he talked to physicists and his father's colleagues about the legacy of Many Worlds.**

THE FATE OF THE UNIVERSE

THE BASICS

The universe will not always look the way it does now. Eventually, stars will burn out, planets will die, and distant galaxies may become forever invisible. Researchers used to think the fate of the universe hinged only on the precise amount of matter and energy in it. If that quantity exceeded a critical density, expansion would reverse and the universe would collapse in a reverse of the Big Bang, called the Big Crunch. If the universe had less than the critical density, it would expand forever as it entered the big chill.

Now we know that 70 percent of the energy density is contributing to a more rapid expansion of space. Everything depends on how dark energy shakes out. Maybe the acceleration will slow down in the future, resulting in a Big Crunch, but there's no reason yet to think so. The most obvious scenario is that the acceleration will continue indefinitely. Anything not bound together by gravity will eventually be pulled apart as the expansion grows stronger. Our galaxy and its neighbors will end up alone in a vast darkness.

After about 100 billion years, we may lose all evidence of the Big Bang. Without any galaxies around, we won't be able to detect the expansion of space. Even the cosmic microwave background photons will become invisible, stretched to wavelengths bigger than our galaxy.

ON THE FRONTIER

Over the next trillion years, galaxies will run out of gas to form new stars. Within 100 trillion years, even the smallest, longest-lived stars around us will have begun to fade. Eventually all that's left will be burned-out white dwarfs, neutron stars, and black holes spiraling into supermassive black holes in galactic cores. Quantum processes may cause black holes to slowly evaporate. In 10^{100} years, the universe will be completely dark, a thin gas of photons and elementary particles. Depressing, no?

COCKTAIL PARTY TIDBITS

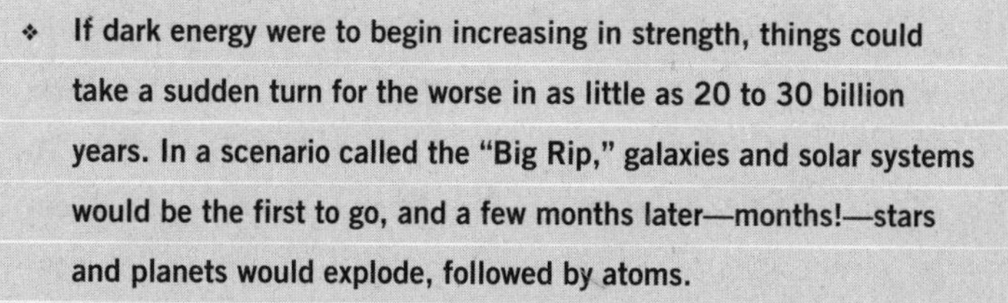

- **If dark energy were to begin increasing in strength, things could take a sudden turn for the worse in as little as 20 to 30 billion years. In a scenario called the "Big Rip," galaxies and solar systems would be the first to go, and a few months later—months!—stars and planets would explode, followed by atoms.**
- **Some researchers think the universe may be cyclic, going through a series of Big Bangs, each followed by a long period of dark energy.**

THE ARROW OF TIME

THE BASICS

Even tougher than forecasting the future is figuring out why there's a future at all. Remember when we talked about entropy we said there's nothing in the laws of physics demanding that time must flow like a river. Instead we perceive the passage of time because everything is winding down from a more-ordered state to a less-ordered one. Eggs fall and break, but they don't reassemble themselves like the T-1000 in *Terminator 2* and jump back up on the counter. That would be time running in reverse.

The same goes for the universe. It just so happens that the Big Bang set the universe up with a bunch of concentrated matter, which gravity could then pull together into stars, galaxies, and the rest of the structure we see today. To scientists, that's weird. Why would all the matter and energy in the universe be set up just so at the moment of the Big Bang?

One step toward an answer comes from inflation, which says the observable universe was once compressed into an exceedingly small region that was driven to inflate by a relatively small lump of vacuum energy. Scientists have proposed that maybe inflation happened because the universe before inflation was undergoing wild quantum fluctuations in energy that were big enough to create a whole universe.

ON THE FRONTIER

Unfortunately, ideas from quantum gravity indicate that even a small glob of inflationary vacuum energy is highly ordered, even more so than the Big Bang.

But there may be hope for time yet. Maybe our hypothetical preinflationary universe was totally empty except for a thin coat of dark energy, the stuff that's driving today's universe to expand ever faster. If the dark energy randomly became a little denser in one patch, then—bang, inflation!

COCKTAIL PARTY TIDBITS

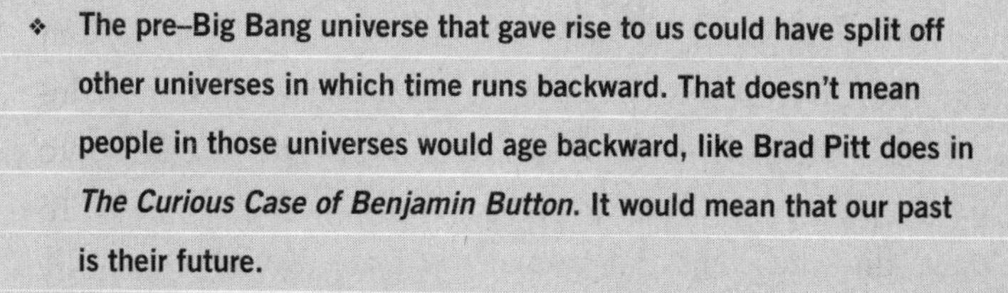

- **The pre–Big Bang universe that gave rise to us could have split off other universes in which time runs backward. That doesn't mean people in those universes would age backward, like Brad Pitt does in *The Curious Case of Benjamin Button.* It would mean that our past is their future.**
- **Did you ever see the episode of *Futurama* where disembodied brains invade Earth? When pondering vacuum fluctuations, researchers worry how frequently the vacuum would vomit up something as complicated as a brain, which they call a Boltzmann brain. If it happened a lot, that would be weird, even to scientists.**

THE MULTIVERSE

THE BASICS

Scientists have had a hard time explaining why the universe is the way it is. In some ways it seems suspiciously neatly set up or "fine-tuned" for the emergence of stars and galaxies. If matter and antimatter were perfectly symmetrical, for example, the universe would have no structure at all. Many scientists would like to find a single equation or set of equations that simultaneously showed why all the funny details of the universe have to be the way they are.

On the other hand, it could be that our universe is an accident of sorts. Things get more interesting if you imagine that our observable universe is only one small patch of a larger "multiverse," similar to what we talked about above. Each universe could have its own physical laws. In ours, the laws allowed stars and galaxies to form, leading to conditions that support planets and life.

Scientists have been drawn to this line of thinking lately by dark energy. Remember that the cosmological constant is really small but not quite zero, which is hard to explain. Right now, the best idea scientists have is to assume that there are many universes that have different dark energies, and we live in one capable of supporting life. This idea is called the anthropic principle.

ON THE FRONTIER

Scientists have proposed that the astronomical number of forms that string theory can take may actually help solve the dark matter problem. Maybe these mathematically possible universes are real and evolve from one to another in a way that will inevitably produce a universe like ours. This idea is called the "anthropic landscape" of string theory. Not everybody likes the anthropic principle, because it means we have to give up the goal of finding that one equation that answers all our questions about the universe. But researchers hope they can use it to come up with some testable predictions.

COCKTAIL PARTY TIDBITS

- **We can't use the fact of our own existence to prove there are multiple universes, for the same reason we can't use it to tell us how common life is outside of Earth. Logically, our existence is equally compatible with a single universe.**
- **Stars might be a common feature of other universes. Researchers have simulated what would happen if the forces of nature were a little different, and it looks as if stars would still form. Let there be light!**